BIODIVERSITY
HERBAL MEDICINE

Edited by

D. Marngar
S. Jyrwa

SYNOD COLLEGE
Shillong - 793002, Meghalaya, India
&
AKANSHA PUBLISHING HOUSE
NEW DELHI-110002 (INDIA)

AKANSHA PUBLISHING HOUSE
4649-B/21, Ansari Road
Darya Ganj, New Delhi - 110 002
Ph.: 23263193 / 9811582579 / 09435116718
Fax: 011-23263193
Email: ektabooks@yahoo.com

Biodiversity: Herbal Medicine

First Published 2009
ISBN 978-81-8370-154-9

[The responsibility for the facts stated, conclusions reached etc., is entirely that of the contributors. The publisher & editors are not responsible for them, whatsoever.]

PRINTED IN INDIA

Published by Akansha Publishing House, New Delhi and Printed at Aryan Enterprises, Delhi.

Contents

Preface

In an effort to contribute towards Biodiversity study, inventorisation, traditional knowledge, data base development, conservation, cultivation and sustainable utilization of medicinal plants in the country, Synod College, Shillong organized a National Seminar "Biodiversity: Herbal Medicine" on the 20th – 21st September, 2007. This book is a compilation of the contributions of various authors from different parts of the country who participated in the seminar. It is sincerely hoped that this compilation would provide necessary information to various scholars and communities about the need of conservation and sustainable use of biodiversity resources.

India is rich in multiplicity of life forms, diversity of ecosystems, species and genetic pool within species. Out of the world's important 'mega-diversity' countries, Eastern Himalayas and Western Ghats are among the 25 'hot spots'. The enormous variation in the vegetation pattern of India is due to mosaic of geo-climatic conditions. India consists of about 49,000 species of plants, which constitutes about 12% of the plant species of the world with high endemism. Northeastern region is rich in cultural, linguistic, ethnic and biodiversity that makes it an important global gene centre.

70-80% of the population in India relies on plants as the only source of medicine; over 7,000 plants out of 8,900 species of ethnobotanical interest are of medicinal value. Herbal plants generate economic value-extraction of medicinal products and other compounds. An estimated 40% of the global economy is based on biological products and processes. It is estimated that an annual turnover of herbal products in India is about $ 63 billion. It is estimated that there are about 25 million cases of cardiovascular diseases, similar number of cases of diabetes, 2.4 million cases of cancer, one million cases of cerebrovascular diseases / strokes and 5.1 million cases of HIV/AIDS in our country. These and many other diseases may increase many folds in the years to come.

In the context of these menaces to the life of human being, there is an urgent need to discover new drugs and the lead on this field would come from the traditional herbal medicines. India has over one million traditional village level healers and several millions of knowledgeable households who are well-versed in traditional home remedies.

The main problem is that 1 out of 8 herbal plants are endangered and the reasons are—unplanned human activities and ruthless exploitation. If over-exploitation continue, it may lead to extinction of plant species and loss of opportunity to discover more useful forms. The rural people in India have always known that plant resources mean their survival.

To meet these problems, challenges and action plan need to follow wisely –

- Conservation, sustainable utilization and management of herbal plants be given top priority.
- The age old traditional knowledge system is needed for integration of indigenous and community knowledge systems related to herbal plants into the mainstream (modern) knowledge systems.
- Clinical trials and toxicological studies for traditional compounds needed.
- Biotechnology to become as an integral part of all aspects of germplasm acquisition, characterization, inventorisation, conservation and herbal plant resource management.
- Bioprospecting of compounds, chemical analysis of plants for new leads on novel active molecules.
- Immediate need to prepare databases on medicinal plants and establishment of patenting cells (offices).
- Safeguards against bio-piracy, establishment of ownership since knowledge of traditional remedies for health cares are still no where in IPR holdings.
- Intellectual Property Rights (IPR) issues and exchange of germplasm need for strict National and States legislation.
- Consortium for collection of raw materials, processing, marketing and export.
- Value addition near the farming areas to reduce the stronghold of the middlemen.
- Programmes for training of farmers or school dropouts on mass multiplication and cultivation of medicinal and aromatic plants.
- Close linkage between universities – research institutions – industries is needed.
- Active involvement of State Governments in enhancing funding for R&D in the area of medicinal plant research with emphasis on the more endangered plants.

March, 2008

D. Marngar
S. Jyrwa

Acknowledgements

This project would never have been accomplished without the encouragement and participation of many people. Our deep appreciation goes to each of the contributors for accepting the challenge of preparing the papers and sharing their ideas. We extend our sincerest thanks to each of them for their contribution.

We extend a heartfelt thanks to all the participants of the National Seminar on "Biodiversity: Herbal Medicine" which was held at Synod College, Shillong, on the 20th – 21st September, 2007.

We are deeply grateful to Dr. D. Wanswett, the Principal of Synod College, Shillong without whom this project would never have been possible. We appreciate his keen interest and tireless effort in making this seminar a success.

We also extend our sincerest appreciation and thanks to Dr. M. Rani, the Coordinator IQAC, Synod College, and at the same time the coordinator of the Seminar, who has coordinated this seminar successfully. We congratulate him for a job well done.

We also thank the North Eastern Council, Shillong and Directorate of Higher and Technical Education, Government of Meghalaya, Shillong for providing financial support towards the organisation of the seminar. Finally, we express our appreciation to all the teachers of Synod College for their participation in making this seminar a success and to all who are involved in making the publication of this volume possible.

List of Contributors

A. J. Gogoi, S.P. Saikia, P.B. Kanjilal and B.S. Bhau	:	Plant Genomics Laboratory, Medicinal, Aromatic & Economic Plant Division, North East Institute of Science & Technology, Jorhat–785 006, Assam.
Akhil Baruah	:	Department of Botany, Darrang College (GU), Tezpur -784001, Assam
Alee Sarma	:	Department of Botany, Dimoria College, Khetri
B.Kharbuli	:	Department of Zoology, North Eastern Hill University, Shillong-793022
C.R. Sahu and J. Ghosal	:	Cell and Developmental Biology Laboratory, Department of Zoology, University of Kalyani, Kalyani – 741235
D. Marngar	:	Department of Zoology, Synod College, Shillong-793002
D.K. Hore and P.K. Singh	:	National Bureau of Plant Genetic Resources, Umiam – 793 103, Meghalaya.
Dibyendu Chakraborty	:	Sikkim Consultancy Centre (a Division of Webcon, Kolkata), Targain's Lee, Church Road, Gangtok, Sikkim–737 101
G. G. Maiti	:	Dept. of Botany, Kalyani University, West Bengal
Jagatpati Tah	:	Genetics and Plant Breeding Laboratory, Department of Botany, Golapbag Campus, Burdwan University, Burdwan – 713104, W. Bengal.

K. Medhi, B.S. Bhau S.P. Saikia	:	Plant Genomic Laboratory, Medicinal, Aromatic & Economic plant Division, North East Institute of Science & Technology, Jorhat – 785006, Assam. E-mail:bsbhau@gmail.com; bhaubs@rrljorhat.res.in
Manashi Kalita	:	Department of Botany, Handique Girls' College, Guwahati
Mandy M. Kharbyngar, Kyntiewlang M. Kharbyngar and **Jacqueline Dkhar**	:	Department of Zoology, Sankardev College, Shillong
Nilakshee Devi	:	Department of Botany, Gauhati University, Guwahati - 781014
N. J. Das	:	Department of Agronomy, Assam Agriculture University, Jorhat, Assam.
Pratibha Sharma, Hemen Sarma, C.M Sarma	:	Department of Botany Gauhati University, Assam E-mail: prativasarma@rediffmail.com hemens02@yahoo.com
Rakshak Kumar and **S. R. Joshi**	:	Department of Biotechnology and Bioinformatics, North Eastern Hill University, Shillong – 793 022.
Ranjan K. Bhagobaty, Pramal Biswa and **S. R. Joshi**	:	Microbiology Laboratory, Department of Biotechnology and Bioinformatics, North Eastern Hill University, Permanent Campus, Umshing, Shillong-793022. Email: rbhagobaty@yahoo.com
R.M. Syiem, A. Chatterjee, S. Majaw and **S. R. Joshi**	:	Department of Biotechnology and Bioinformatics North Eastern Hill University, Shillong-793 022 Email: rymphang@yahoo.com
S. Majaw and **D. Syiem.**	:	Department of Biochemistry, North Eastern Hill University, Shillong 793 022, Meghalaya
S. P. Saikia, B. S. Bhau, Kalyani Medhi and **P. B. Kanjilal**	:	Division of Medicinal, Aromatic & Economic Plants, North East Institute of Science & Technology (Erstwhile Regional Research Laboratory), Jorhat – 785 006, Assam

Sristisri Upadhyaya and **L.R.Saikia**	:	Department of Life Sciences, Dibrugarh University. Dibrugarh-786004 E-mail:lrsdueco@rediffmail.com
S. Sarkar	:	Department of Botany, Rangia College, Rangia – 781 354, Kamrup, Assam.
Sumita Kharmawphlang	:	Ayurvedic Physician, Civil Hospital, Nongpoh, Meghalaya

1

Inventorisation and Status of Medicinal Plants Diversity with Reference to Northeastern India

D.K. Hore and P.K. Singh

1. Introduction

The varied physiography of Northeastern region together with rich vegetation cover and floristic diversities has made the region a 'HOT SPOT' of the country. The region comprises of 8 states which is supposed to be a contiguous belt of Eastern Himalaya where the altitudes ranges between 45 to 7090 m; rainfall ranges between 300 to 12000 mm annually and the temperature usually between 0° to 36° C. Thus, the environment prevails within the region in between Tropical to Alpine climate. The region is a part of Eastern Himalayas where the original primary forest habitat comprises at least 9000 plant species, out of which 39% plant species are endemic. Medicinal plants are also an integral part of these floristic components, which are frequently used in Indian system of Medicines like Homoeopathy and Ayurveda. It was estimated that 43% of total Indian flora exist in this region, out of which at least 70% of the medicinal plant species are available in forested area. Harvesting of medicinal plants is normally constituted hardly 5% of the total forested area. In these habitats, at least 10% of the total species belong to medicinal plants category.

The region is also dominated by the tribal populations. At least 120 tribes / sub-tribes are dispersed all over the region. Health care facilities in the remote areas is practically absent. Most of the inhabitants depend on the plant products for medicinal treatment. At least, there are about 400 such plant species, which are utilized to cure different diseases and ailments (Table 1) in the region.

Table 1

Number of Medicinal plant species utilized against different diseases and ailments

Uses	No. of Species	Uses	No. of Species
Anti-fertility	18	Blood Cholesterol	16
Stress control	17	Anti-bacterial	26
Cardiovascular	17	Anti-viral	11
Hypertension	21	Anti-protozoal	13
Blood sugar control	25	Anti-fungal	09
Respiratory disease	33	Galactogogue	10
Liver function	33	Anti-ageing	10
Gastro-intestinal disorder	40	Analgesic	18
Urinogenital disorder	26	Vitality enhancer	30
Anticancer	21		

Adaptation and ecological environment is a prime factor to grow any medicinal plant species. Based on the altitude and ecological zonation the region has been classified as Tropical, Sub-tropical, Temperate and Alpine. Accordingly, an inventorisation of various important medicinal plant species was conducted.

The list was strictly in context to the region.

Tropical Region

Adhatoda vasica, Andrographis paniculata, Aquallaria agallocha, Centella asiatica, Dioscorea alata, Ocimum basilicum, Plantago erosa, Withania somnifera, Vitex negundo, Paedaria foetida, Cryptolepis buchanani and *Acorus calamus.*

Sub-tropical Region

Artemisia maritime, Hydnocarpus kurzii, Lavendula vera, Litsea cubeba, Mucuna pruriens, Pogostemon cablin, Zanthoxylum armatum, Curcuma caesia and *Leonotis nepetaefolia.*

Temperate Region

Coptis teeta, Geranium nepalensis, Panax pseudoginseng, Swertia chirata, Picrorrhiza kurroa, Satyrium nepalensis, Rubia cordifolia and *Taxus baccata.*

Alpine Region

Aconitum ferox, Aconitum heterophyllum, Illicium griffithii, Berberis spp., *Podophyllum hexandrum, Rheum emodi* and *Delphinium subulatum.*

2. Priority Species

The occurrence of species population in the region and in context of disease control, a list of priority species has been inventorised and drawn up here. This will help to envisage the future policy strategy for sustainable development of both the species as well as the region.

Table 2
Priority medicinal plant species of the region

Species	Habitat	Used against the ailment
Panax pseudoginseng Wall.	Hill forest	AIDS
Albizia procera Benth.	Forest	Cancer
Alisma aquatica L.	Stagnant water body	Cancer
Ophiorrhiza mungos L.	Hill forest floor	Cancer
Artemisia maritime L.	Forest edges	Malaria
Plantago erosa Wall.	Forest edges	Malaria
Dichroa febrifuga Lour.	Hills	Malaria
Dioscorea alata L.	Plains	Leprosy
Hydnocarpus kurzii (King) Warp.	Sub-tropical forest	Leprosy
Averrhoa carambola L.	Assam plains	Jaundice
Bixa orellana L.	Sub-tropical forest	Jaundice
Bridelia stipularis Bl.	Plains	Jaundice
Hedyotis auriculata L.	Plains	Jaundice
Hygrophila salicifolia Nees	Plains	Jaundice
Paederia foetida L.	Forest areas	Paralysis
Caladium bicolor Vent.	Plains & Hills	Paralysis
Mitragyna rotundifolia (Roxb.) O.K.	Plains & Hills	Paralysis
Litsea cubeba (Lour.) Pers.	Plains & Foothills	Paralysis
Kyllingia nemoralis (J.R. & G. Frost.) Dandy ex Hutch. & Dalzel.	Plains	Diabetes

(Contd...)

Species	Habitat	Used against the ailment
Emilia sonchifolia Dc.	Plains & Hills	Night blindness
Symplocos racemosa Roxb.	Lower hills	Night blindness
Trewia nudiflora L.	Plains	Gout
Randia spinosa Bl.	Plains	Abortifacients
Picrorrhiza kurrooa Royle	Hill forest	Jaundice
Potentilla sundaicaI (Bl.) O.K.	Rocky hills floor	Antidote for snakebite
Clerodendron colebrookianum Walp.	Tidal place	Heart diseases
Swertia chirata Buch.-Ham.	Hills	Stomach disorder
Coptis teeta Wall.	Temperate hills	Malaria
Mucuna pruriens Hook.f.	Wasteland	Neurological diseases
Psychotria ipecacunha Stockes	Subtropical	Psychological disorders
Campylandra aurantiaca Baker.	Temperate hills	Psychological disorders
Aconitum ferox Wall. ex Ser.	Temperate hills	Gastric disorder
Houttuynia cordata Thunb.	Hills	Dysentery
Vitex trifolia L.	Plains	Tuberculosis
Thevetia peruviane (Pers.) Merz.	Hills & Plains	Abortifacients
Barleria cristata L.	Hills	Abortifacients
Berginia ciliate (How.) Sternb.	Hills	Abortifacients
Phytolacea acinosa Roxb.	Hills	Rheumatism

3. Scope and Reasons of Cultivation of Medicinal Plants

There are about 80% people from developing countries that depend on herbal medicine. 25% of the medicines are derived from various plants (Fransworth & Soejarto, 1991). In this venture, 70 billion US Dollar is involved annually in world trade. Hence, there is 7% growth rate per annum (Gera *et al.* 2003). Enough opportunity would be existed, if systematic attempts for marketing and cultivation of medicinal plant species were adopted. This will facilitate the following approaches towards the sustainable development of the region.

- To ease pressure on natural resources.
- To make available fresh, genuine and quality raw material for manufacturing standardized and efficacious drugs.
- To evolve better strains and high yielding crops of medicinal plants through improvement programme and tissue culture techniques.
- To check the use of spurious substitutes and adulterants which have resulted in the deterioration of the standard of drugs of ISM.
- To standardize the collection, storage and post-harvest technology for medicinal plants.
- To provide regular and alternative source of income to the farmers for amelioration of their economic conditions.
- Introduction and domestication of useful exotic drug plants to minimize import and maximum export.
- To conserve the biological and genetic diversity in medicinal plants for the posterity.

Propagation Mode of Medicinal Plants

The mode of propagation and multiplication of the medicinal plants can be done seasonally. The agro-technique should be developed first where it is unknown. In case of many medicinal plants, the desired agro-technique is still to be developed, based on the relevant agro-climatic and soil conditions.

Depending on the mode of multiplication, there are three types of procedures through which the desired material can be multiplied in the field. In many cases, both the seeds and cuttings are employed for such propagation.

- Propagation through seed.
- Propagation through cuttings.
- Propagation through specialized structures.

Examples –

A. **Propagation through seeds :** *Altizia procera* Benth, *Ophiorhiza mungos* L., *Artemisia maritime* L., *Plantago erosa* Wall., *Bixa orellana* L., *Bridelia stipularis* Bl., *Picrorhiza kurrooa* Roxb., *Taxus baccata* Hook.f.

B. **Propagation through cuttings:** *Dichroa febrifuga* Lour., *Artemisia maritime* L., *Cryptolepis buchanani* Roem. & Schult.

C Propagated through the specialized structures:

Species	Plant parts
• *Alisma aquatica* L.	- Suckers
• *Caladium bicolor* Vent.	- Suckers
• *Kyllinga nemoralis* J.R. & G. Frost	- Suckers
• *Potentilla sudaica* (Bl.) O.K.	- Roots
• *Homalomena aromatica*	- Rooted suckers
• *Alpinia galangal* Willd.	- Rhizome
• *Coptis teeta* Wall.	- Rhizome
• *Dioscorea alata* L.	- Suckers
• *Campylandra aurantiaca* Baker	- Bulb

4. Conservation Strategies

Depending upon the mode of propagation of the species, the following practices can be adopted for the conservation of medicinal plants.

i) ***In-situ* Conservation:** This procedure is already in practice in the region in the form of Biosphere Reserve, Sacred grove, Gene sanctuary and habitat conservation. Species like *Taxus baccata, Coptis teeta, Nepenthes khasiana, Aconitum ferox* are already conserved in different habitats. The reason being the population is very limited within the habitat. Moreover, unless it is protected, there will be rampant extraction of raw material from these habitats.

ii) ***Ex-situ* Conservation:** This is normally maintained in MTS/LTS module, if it is seeded one. Regulation of temperature and moisture is again dependent on the species. If the material is vegetatively propagated, in that case, it is conserved and maintained in the field gene bank. Time to time care is required in order to maintain the population of the species. Cryopreservation is an alternative method of *ex-situ* conservation, where the seed samples are endangered or endemic. These are normally preserved in order to maintain the viability. Most of the recalcitrant seeds are conserved in this method. In this case, fine seed grains/spores are normally stored in sample vials and placed in cryo-tank. The temperature of cryo-tank is maintained under -196°C, with the help of liquid Nitrogen pressure and the seed viability remains more than 100 years.

iii) ***In-vitro* Conservation:** Vegetatively propagated material can easily be multiplied through this technique, subject to availability of suitable protocol. A large number of plantlets can be regenerated through this system. However, the maintenance cost of the experiment is very expensive.

5. Marketing Constraints

In context of NE region, there is no large scale cultivation practice of medicinal plants; the basic reason is no marketing opportunity. Raw materials at least can be grown provided if there is assured market or buy back policy. Whatever, material collected from nature does not often suit the industries, because of the quality of the material. Hence there is a big loss; this is classified in two ways –

(i) Chances of extinction of species due to non-judicious extraction

(ii) Unawareness about the physiological maturity and effectiveness of desired compounds of the material. The existing trade and marketing bottlenecks are due to –

- Lack of regulated market for raw materials.
- Middlemen and pharmaceutical industries keep hold on the market price.
- Random extraction from the wild habitat, resulted declines of market price on one hand and the survival of the species population on the other.
- Unless an MOU or buy back policy is introduced at Government level, the situation will always remain out of control. Contract farming system may also be adopted.

6. Future Strategies

In order to counterbalance the marketing situation for the growers, entrepreneurs and buyers the following systematic approaches are required –

- Identification of market demand species to be cultivated for raw materials.
- Agro-technique of the species to be developed and then cultivated.
- Determination of physiological maturity of the material/plant parts.
- Harvesting procedure, quality standardization, processing etc. as per the requirement of market (i.e. value addition).
- Settlement of terms and conditions of MOU for the item(s), with respect to its marketing.

In this venture, the involvement of Scientists, Farmers, Government, Entrepreneurs and the buyers are absolutely necessary. In many cases, patent law, IPR issues, traditional knowledge, the question of benefit sharing and issues related to biopiracy may arise. With their strong support and action the system can only remain viable.

REFERENCES

Fransworth N.R. and Soejarto D.D. 1991. Global importance of medicinal plants, *Conservation of Medicinal Plants* (Eds. Akerale, D., V.H. Heywood and H. Synoge). Cambridge University Press, Cambridge, U.K.

Gera M., Bisht N.S. and Rana A.K. 2003. Market information system for sustainable management of medicinal plants. *Indian for.* 129(1):102-108.

2

Medicinal Aromatic Plants of West Bengal: A Profile

G. G. Maiti

1. Introduction

West Bengal enjoys a unique landmark from the floristic point of view and it is the only state of India where its geographical location is endowed with a flora ranging from littoral flora of Sundarbans to the Himalayan hilly temperate and alpine zone of the Darjeeling districts. The total geographical area is 87,676 sq. km. i. e. 2.7% of the total geographical area of India (Bhattacharyya, 1997).

The state can be divided into two natural geographical regions, the Himalayan region and the Gangetic Plain. However, the agro - ecological zones are as follows: (i) Himalayan zone, (ii) Sub-Himalayan zone, (iii) Alluvial zone, (iv) Lateritic zone and (v) Coastal zone.

Prain (1903) had specified the flora under four botanical divisions as also considered by Basak & Guha Bakshi (1977): (1) Western part of West Bengal is lying between the river Bhagirathi- Hoogly and the Chhotanagpur Plateau, (2) a part of the North Bengal covering the land north of Bhagirathi, (3) the strip of land lying east of Bhagirathi and (4) the Western part of Sundarbans mainly occupied by the districts of South 24 Parganas. Further, the mountainous district of Darjeeling, Jalpaiguri and Koochbihar can be added as a sub-Himalayan zone.

Due to the favourable agro-climatic conditions, West Bengal has always been recognized as one of the most populated regions of the country and nearly 72% of the population is in rural areas which spread to over 43,000 villages. The population has reached 88.75 millions with density of population standing 400 per sq. km. Due to steady increase in density of population, the state is under a severe threat from land use pattern and eventually the land and water bodies supporting natural vegetation are being brought under human habitation and related activities. About 65% of the total area is used for agricultural purposes and nearly 95% of the rural people of West Bengal are engaged in agriculture.

Floral Status of West Bengal

The floral status of West Bengal is most impressive in Terai, Duars, Darjeeling Himalayan region and in the mangrove forests of Sundarbans. The vascular plants as estimated from the relevant literature reveals that the Flora of West Bengal includes 450 species of pteridophytes, 15 species of gymnosperms and 3,580 species of angiosperms out of the Indian estimate. In other way, it is estimated that in West Bengal there are about 1,280 herbs, 400 shrubs, 250 climbers of both woody and herbaceous, about 150 trees and some parasites, carnivorous plants etc., along with about 30 species of aquatic herbs.

Status of Medicinal and Aromatic Plants

The occurrence of medicinal and aromatic plants in West Bengal is significant. There are many important medicinal trees such as *Azadirachta indica, Terminalia cuneata, T. belerica, T. tomentosa, T. chebula, Holarrhena pubescens, Emblica officinalis, Soraca asoca, Anthocephalus chinensis, Alstonia scholaris, Butea frondosa, Aegle marmelos, Cassia fistula, Crataeva magna, Madhuca langifolia, Nyctanthes arbortristis, Vitex negundo* and *Strychnos nux-vomica.*

There are nearly 40 species of aquatic herbs and 35 species of wasteland herbs. The common medicinal and aromatic herbs are *Acalypha indica, Achyranthes aspera, Acorus calamus, Aloe vera, Amaranthus spinosus, Andrographis paniculata, Bacopa monnieri, Catharanthus rosens, Centella asiatica, Costus specious, Curcuma amada, Hygrophyla spinosa, Kaempferia galangal, Ocimum tenuiflorum, O. basilicum, O. gratissium, Psorolia corylifolia, Uraria picta, Withania sommifera, Zingiber officinale,* etc. The shrubby members are *Ambroma ungusta, Artemisia nilagarica, Berleria prionitis, Calotropis gigantean, Plumbago zeylanica, Adhatoda zeylanica, Phlogacanthus thyrsiflorus, Lawsonia innermis, Woodfurdia fruticosa,* etc. A few climbers can be mentioned as *Abrus precatorius, Tinospora cordifolia, T. crispa, Piper nigrum, Argyreia nervosa, Gymnema sylvestre, Tylophora indica, Aristolochia indica, Cissampelos pareira, Asparagus racemosus, Gloriosa superba, Cocculus hirsutus, Coccinea grandis* etc. Dicotyledonous members are very much used as medicinal and aromatic plants than the pteridophytes and lastly the monocotyledonous plants. Similarly, herbaceous plants are much used at 41% compared to shrubs 26%, trees 17% and climbers 16%. However, there are many marsh and aquatic plants also (Gupta & Singh, 2005).

2. Historical Review of Medicinal and Aromatic Plants

Keeping in mind it can be mentioned that the Rig.Veda (4500-1600 B.C.), an old repository of human knowledge, had quoted 67 plants of medicinal value. Artha Veda and Yajurveda had mentioned 200 and 81 species respectively that have the therapeutic value. Charak (1000 B.C.) and Sushruta (800 B.C) had quoted 341 and 395 medicinal plant species respectively in their Samhitas. Now, the ethno botanical studies have further added to about 1,680 species of medicinal value (Bodding, 1925, Pal & Jain, 1998). In India as well as in West Bengal the uses of herbal drugs are reported to utilize nearly 450 plant species in different formulations (Prajapati *et al.* 2003). The number has increased by 12-15 % up till now.

In the last 60 years, the basic active principles like alkaloid, glycosides, steroid compounds and essential oils etc. have been isolated from higher groups of plants. 2,000 alkaloids were known to the world during the year 1950 and by 1971 the number was increased to 4,000 and it is estimated to be about 10,000 alkaloids in 2010 (Hazra *et al.* 2005).

Very recently the steroidal sapoginous substances have also attracted phytochemists to give more attention to their researches. The plant based drug industries have four major segments in using the plants species. These are - (1) Plant drugs for WHO recognized Indian System of Medicine, (2) non-prescription items involving plant parts, extracts and chemicals – used over the country, (3) essential oils and (4) phyto-pharmaceuticals (WHO, 2000). Besides these, there are many derived plant materials used such as fats, oils, waxes, latex, pectins, resins, gums, oleoresins and other exudates, vegetable dyes, tannins and other bio-chemical compounds including the hallucinogenic substances (Schultes, 1963). Plants with active principles of greater complexity and novelty are of much demand in the speedy expansion of the global phyto-pharmaceutical industries.

In West Bengal there are nearly 642 licensed pharmaceuticals (Honda, 1998) and besides this there are many small manufacturing units and thousands of Vaidyas, Kaviraj and Kakims who prepare their own drugs from various medicinal plants. Moreover, medicines used in the Indian medicinal system, herbal cosmetics, herbal nutrients etc. are all the subjects of industries. The direct utilization of plants and plant products as raw materials for herbal formulation, preparation of decoction, tinctures etc. have much increased. Similarly, production of essential oils, oleoresins, curcumins, perfumes etc. have increased in a number of industrial companies (Maiti, 2002; Basak, 2003; Lewag & Foster, 1996).

3. Market of Medicinal and Aromatic Plants

India with its vast medicinal plant diversity has the potential to hold a key position in the international market of herbal medicine. It has to cope with the competition from leading Asian countries like China, Korea and other developed countries of Europe and United States. Compared to China, where annual foreign exchange earning by way of export of medicinal and aromatic plants touches Rs.15,000 crores accounting for 30 percent of the global herbal market, India's earning from the same is a meagre Rs.5,560 crores (Koetlar, 2003). Currently there is an increasing concern regarding the quality of the medicinal plant materials being delivered to local pharmaceutical industries or exported to foreign countries. The market of herbal medicines in developed countries especially in Europe and United States is highly regulated. According to a survey, India is one of the major exporters of crude drugs to six developed countries – USA, Germany, France, Switzerland, UK and Japan, who share among them 75-80 percent of the total export market (Chatterjee 2002, Anonymous, 2003).

In West Bengal there are only two markets for medicinal plants, one at Kolkata (near Burrabazar-Chitpur area) and the other at Siliguri but the situation is not well enough compared with that of Mumbai and Delhi. There are several reasons regarding the dubious trading of medicinal and aromatic plants in both these places. However, in 2000, the medicinal plant trading companies of Kolkata have formed an association in the name of West Bengal Herb and Crude Drug Dealer's Association (WBHCDDA) to look after the interest of the community and trade of medicinal plants as a whole.

The Indian market for export of herbal products is promising (Table 1). The export is mostly to Europe such Germany, France, U K, Spain and also to USA Japan, Australia, Taiwan etc. (Pramanik, 2005). The major medicinal plant products exported from India during 2002-2003 are presented in Table – 2.

Prioritized Identified Medicinal Plants

The National Medicinal Plant Board (NMPB) has prioritized 32 medicinal plants for West Bengal (Table – 3) on the basis of their demand in domestic and international markets. It is expected that programmes on cultivation of selected plants, location wise in different regions of Bengal will be implemented. In implementation of proposed plans for cultivation at different agro-climate zones, several organizations have already set up for organized cultivation of medicinal and aromatic plants. A few can be mentioned here as Bidhan Chandra Krishi Viswavidyalaya, Ramakrishna Mission Ashrama, Narendrapur, Burdwan University, North Bengal University, Agri–Horticural Society of India, Alipur, Kolkata, Ramakrishna-Vivekananda Ashrama, Panihiti. However, Central Institute of Medicinal and Aromatic Plants (CIMAP), Lucknow has developed large number of Medicinal and Aromatic Plants including some superior varieties (Khanuha *et al.* 2007).

Drawbacks in Marketing of Medicinal Plants

The major concern areas in the business of plant materials are- (i) unscientific harvesting, (ii) unscientific production of raw materials, (iii) unavailability of cultivated lands, (iv) unorganized cultivation, (v) fluctuation in demand and supply, (vi) poor system of quality control, (vii) lack of availability and supply of superior materials, (viii) poor marketing infrastructure, and (ix) lack of coordination among different stakeholders. All these concerns need to be shared by the farmers, traders, sellers, resellers including major retailers, pharmaceutical companies, NGO, wholesalers, exporters etc. in the entire system of business of the medicinal plant species.

4. Situation of Medicinal Plants in Nature

India exhibits a wide range of climatic and topographic variations embracing 8% of the estimated biodiversity of the world. Out of the total of 3,500 medicinally important flowering plants of tropical India, representing about 40% of the total flowering plants in this region, 300 are the major medicinally important species. But due to regular collection from nature such as fields and forests these plants are becoming rare and endangered. They are being regularly over exploited and are on the verge of extinction (Puspangadan, 1995). The declining of diversity of medicinal and aromatic plants in the field is a concern to save them in nature. Thus it needs regular protection, maintenance and conservation in nature. On the other hand the organized cultivation should not disturb their natural habitat. A few examples can be stated with regards to the over-exploitation of these medicinal and aromatic plants as *Swertia chiirayita, Rauvolfia serpentina, Gloriosa superba, Ipomoea digitata, Dioscorea deltoidea, D. alata, Costus speciosus* etc.

5. Conclusion

Practically, there is no proper marketing facility and there is a lack of supply facilities directly to buyers, retailers, seller and exporters in West Bengal. The demand and supply do not often correlate and thus the rate or cost of the medicinal plants is not fixed. There are also several factors about the identity of actual species and good germplasms or propagules. Furthermore, there are other constrains as the absence of the guidelines for proper cultivation practices, harvesting and post harvesting management, control measure of diseases, storing, packing, etc. Besides these, there is no development or proper role from companies to farmers in contract farming. Thus, there is no success story for corporate farming, cooperative farming or group farming etc. On the other hand, it is significant that the trading rate is recommendable for contract farming of some medicinal and aromatic plants such as *Aloe vera, Andrographis paniculata, Asparagus racemosus, Chlorophytum borivilianum, Coleus forskholii, Emblica offocinalis, Gloriosa superba,*

Gymnema sylvestre, Ocimum tenuiflorum, Piper longum, Plantago ovata, Rauvolfia serpentina, Solanum nigrum, Tinospora cordifolia, Withania somnifera etc.

Table 1
Export scenario of herbal products

Year	Value (Rs. Crores)
1995-1996	13840.00
1996-1997	16199.53
1997-1998	15732.86
1998-1999	16808.70
1999-2000	21590.72
2000-2001	23150.50

Source: MAIRDC. World Trade Centre, Mumbai (2002)

Table 2
Major medicinal plants products exported from India (2002-2003)

Item	Value (US $ million)
Psyllium husk	35.49
Sap and extracts of opium	18.72
Gambodge extract	11.23
Other extracts	7.77
Henna powder	4.27
Ayurvedic and Unani herbs	5.03
Other crude drugs	5.82
Senna leaves and pods	7.63
Sandal wood chips and dust	6.53
Karya gum	2.73
Total	**105.22**

Source: India Trades (2004)

Table 3
32 Prioritized plants and their annual demand

Sl. No.	Name of species	Trade name	Demand (in tones) 2001-'02	2004-'05
1.	*Emblica officinalis*	Amla	22729.5	41782.9
2.	*Saraca asoca*	Ashok	7051.3	10724.2
3.	*Withana somnifera*	Ashwagandha	7028.7	9127.5
4.	*Aconitum heterophyllum*	Atis	270.1	448.4
5.	*Aegle marmelos*	Bael	5381.2	7084.5
6.	*Phyllanthus fraternus*	Bhumi amlaki	2212.6	2985.3
7.	*Bacopa monnieri*	Brahmi	3822.5	6621.8
8.	*Santalum album*	Chandan	635.2	1073.1
9.	*Sweritia chirayita*	Chirata	965.2	1284.7
10.	*Berberis aristata*	Daru haridra	1187.3	1829.4
11.	*Tinospora cordifolia*	Giloe/Gulancha tala	2258.3	2932.6
12.	*Gymnema sylvestre*	Gurhmar	N.A	N.A
13.	*Commiphora mukul*	Guggal	1505.0	2548.9
14.	*Plantago ovata*	Isabgul	N.A	N.A
15.	*Nardostachys jatamansi*	Jatamansi	674.9	866.8
16.	*Solanum surattense*	Kantikari	65.4	100.5
17.	*Andrographis paniculata*	Kalmegh	2005.0	2197.3
18.	*Crocus sativus*	Kesar	N.A	N.A
19.	*Garcinia indica*	Kokum	N.A	N.A
20.	*Saussurea costus*	Kuth	1414.1	1826.3
21.	*Picrorhiza kurroa*	Kutki	220.3	317.0
22.	*Solanum nigrum*	Makon/kok machi	2077.9	2192.2
23.	*Glycyrrhiza glabra*	Mulethi/ Jasthimadhu	873.4	1359.8
24.	*Bryophyllum pinnatum*	Patharchur	37.8	60.8
25.	*Piper longum*	Pippal	3992.5	6280.4
26.	*Chlorophytum borivilianum*	Safed mushli	N.A	N.A
27.	*Rauvolfia serpentina*	Sarpagandha	423.6	588.7
28.	*Cassia angustifolia*	Senna	6462.5	11677.3
29.	*Asparagus racemosus*	Shatavari	10924.7	16658.5
30.	*Ocimum tenuiflorum*	Tulsi	3296.8	5402.9
31.	*Acorus calamus*	Vatsnabh, bach	322.3	3426.8
32.	*Embelia ribes*	Viavidang, biranga	N.A	N.A

REFERENCES

Annonymous, 2003. *Medicinal Plants in India.* Institute of Economic and Market Research, New Delhi.

Basak R. K. and Guha Bakshi D. N. 1977. Floristic studies on the Lower Ganga Plain of the State of West Bengal: Present status, future strategies. *Bull. Bot. Surv. India* 19: 42-45.

Basak S. L. 2003. *Nutritional and Medicinal Values of Plants Foods.* Naya Udyog, Kolkata.

Bhattacharyya U.C. 1997. *Flora of West Bengal.* Vol. 1. (ed. Hajra R. K.) Botanical Survey of India, Calcutta.

Bodding P.O. 1925. Studies in Santal Medicines and connected Folklore. Part I and II *Mem. Asiat. Soc. Bengal.* 10: 1-427.

Chatterjee S. K. 2001. *Cultivation of Medicinal Plants in India.* World Conference Medical and Aromatic Plants. Budapest, Hungary.

Gupta Veena and Singh A. K. 2005. Medicinal and Aromatic Plants: Possible Conversation Approaches. In Gupta S. K. and Mitra B. R. (eds.), *National Seminar on Herbs in Health Care-Information Update.* Ramakrishna Mission Ashrama, Narendrapur, Kolkata. Pp. 50-61.

Handa S. S. 1998. *Quality Control of Medicinal Plants and Plant Products.* CDRI, Lucknow.

Hazra A. K., Banerjee S., Seal T. and Mukherjee B. 2005. Prospect of Secondary Metabolites in Modern Medicine. In Gupta S. K. and Mitra B. R. (eds.), *National Seminar on Indian System of Medicine "Herbs in Health Care. Information Update."* Ramakrishna Mission Ashrama, Narendrapur, Kolkata. pp. 150-160.

Khanuja S. P., Kalra S. A. and Singh A. K. 2007. Need for a strategic approach. *The Hindu-Survey of India Agriculture.* 2007. pp. 186-192.

Koetlar P. 2003. *Marketing Management,* Prentice Hall of India.

Leung A. and Foster S. 1996. *Encyclopaedia of Common Natural Ingredients Used in Food, Drugs and Cosmetics.* John Willey and Sons. INC, New Delhi.

Maiti G. G. 2002. Medicinal uses of some exotic plants in West Bengal. In Maiti G. G., Mahapatra A. K. and Guha Bakshi D. N. (eds.) *National Seminar and Workshop on Indian System of Medicine and Homoeopathy.* Ramakrishna Mission Ashrama, Narendrapur, Kolkata. pp. 62-65.

Pal D. C. and Jain S. K. 1998. *Tribal Medicine.* Naya Prakash, Calcutta.

Prain D. 1903, *Bengal Plants.* 2 Vols. Botanical Survey of India. Calcutta (Rep. Ed. 1963).

Prajapati N. D., Purohit S. S., Sharma A. K. and Kumar T. 2003. *A Handbook of Medicinal Plants: A Complete Source Book.* Agrobios (India). Johdpur.

Pramanik P. 2005. Export of medical plants and herbal drugs from India. In Gupta S. K. and Mitra B. R. (eds.). *National Seminar on Herbs in Health Care Information Update.* Ramakrishna Mission Ashrama, Narendrapur, Kolkata. pp. 178-185.

Pushpangadan P. 1995. *Ethnobiology in India: A Status Report*. Ministry of Environment and Forest, Govt. of India, New Delhi.

Schultes R. E. 1963. The widening panorama in *Medicinal Botany Rhodora* 65: 97- 120.

Sinha R. K. and Sinha S. 2001. *Ethnobotany*. Surabhi Publications, Jaipur.

WHO 2000. *Monograph on Selected Medicinal Plants*. Vol. I. WHO, Geneva.

3

Database on Medicinal Plants and its Public Utilization

D. Marngar and B. Kharbuli

1. Introduction

Today's human population explosion is threatening not only native plants, animals, and their habitats, but also a loss in indigenous knowledge of medicinal and aromatic plants. These changes have left many populations with little or no medical coverage.

Indigenous knowledge is of great potential practical value. It can provide an important information base for local resources management, especially in the Northeast, where conventionally-used data are usually scarce to non-existent, as well as providing a shortcut to pinpoint essential scientific research needs. To be useful for resources management, however, it must be systematically collected and scientifically verified, before being blended with complementary information derived from Western-based sciences.

Local knowledge is a "system of concepts, beliefs and ways of learning", generated and transformed by communities through time, via a systematic process of observation, experimentation and adaptation. It includes a system of classification, a set of empirical observations about the local environment, the development and use of appropriate technologies and a network of social arrangements governing resource use. Through local knowledge, people understand and cope with their particular bio-physical and socio-economic environments, and thus improve the quality of their lives.

Therefore, documentation of indigenous knowledge and history using multimedia has become an extremely important tool to ensure the survival and self-sustainability of indigenous tribes and cultures, to provide evidence of past injustices and to support claims of original ownership. Hence, there is an urgent need for setting up an information system for collection, processing, storage, retrieval and dissemination of information on Indigenous and traditional knowledge on medicinal and aromatic plants in Northeast India. Creation of a computerized database with Indexing and abstracting services facilitates retrospective and current searches.

2. Creation of Database

Medicinal Plant databases will contain actual research data. The Plant Record Type database provides a complete physical and taxonomic description of the plant. The Indication/Preparation/ Administration records would form information on the ailment treated, the plant part used, method of preparation, and how the medicine is given. The Marketing Record Type database relays information on the commercial aspects of the plant such as tariff, patent, and availability information. Data are stored and maintained using software. A separate retrieval software program was written to provide easy data access.

Traditional Medicine (TM) database's content is derived from knowledge of traditional medicine. Its aim is to integrate the ancient wisdom of oriental tradition with modern science and technology.

TM offers three major information databases:

- A prescription database, which provides information on efficacy, dosages, adverse effects, and other knowledge about thousands of traditional drugs.
- A chemistry database, which includes thousands of natural constituents found in plants, showing the chemical formula, chemical structure, and various analytical data, such as spectral data.
- A photo-image database, which shows full-color images of medicinal plants and herbs identified by systematic botanists.

TM contains information useful to researchers searching for new drugs derived from natural products and traditional medicines as -

- To build up information resources, development of databases in the area of MAP's.
- To continuously access information requirements and provide solutions to community of users.
- To evolve and implement programme on education of users and training of R&D scientists.
- To develop, support and enhance public information resources for biotechnology.

3. Indigineous Plants Database

Northeast region of India with its diverse flora is inhabited by numerous ethnic groups with their distinct way of life, dialects believed tradition and cultural heritage offers tremendous scope for ethnobotanical studies but considering the area and the number of tribes inhabiting these densely wooded regions, the studies conducted so far are very meager. With further intensive ethnobotanical study, especially in remote areas, many more of these plants can be discovered. But with the disappearing of forest due to human interference, a numbers of these species have already become scared in the area, and may be extinct. Therefore, it is all the more necessary to carry out extensive ethnobotanical studies to dig out the hidden information about plants and bring such plant under cultivation, if found efficacious after pharmacological and clinical test, before the species are lost forever.

It is true that most of the tribal do not like to part with the information about plants to an outsider but scientists belonging to these specific tribes can easily gather very liable information through direct conversation in the old dialect with their own tribe.

Cultivation of MPs has to be encouraged for meeting the future needs. The cultivation of these plants could provide farmers with an attractive opportunity to increase their income. Some MPs have been selected for propagation and trade. These are:

Gaultheria fragrantissima (Family: Ericaceae) or the Indian Wintergreen is a much-branched, evergreen, aromatic shrub, 1-3 m height, with orange-brown bark, commonly found in the Central and Eastern Himalayas, Khasi Hills and the hills of Western Ghats at altitudes of 1500-2500 msl. It is also grown as an ornamental plant in the hill areas.

The leaves has been known to contain an oil, called as oil of wintergreen, the main use is as an anti-rheumatic, anti-sciatica, painkiller, stimulant, carminative, antiseptic, vermicide and as ingredient in perfumery, insecticidal repellents, soft drinks and dentifrices. The principal constituent of the oil is methyl salicylate. Indian wintergreen oil is distilled in small scale in the Nilgiris and Assam. Methyl salicylate obtained from natural sources is in very high demand in the market as starting material for aspirin and other medicinal compounds. A 100g of leaves yield approximately 0.6gm oil of wintergreen.

Gaultheria fragrantissima

Litsea citrata* syn. *cubeba Blume (Family Lauraceae) is a common deciduous shrub or a small tree found in the Eastern Himalayas, up to an altitude of 850-1000 msl. Flowers and fruits are aromatic and yield on distillation volatile oils. The oil is also used as insect repellent, in perfumery industries and as a basic ingredient in toiletries. The matured fruits on distillation yield about 95% citral, the flowers about 37% citral and the seeds about 20% of pale yellow fat. The leaves and bark of the tree also yields volatile oils and two alkaloids namely laurotetanine and methyl laurotetanine.

Litsea citrata

***Swertia chirata* (Chiretta),** Family: Gentianaceae, is also known as Brown or White Chiretta or Chirata, an erect annual plant of 60-125 cm. tall, found in the temperate Himalayas at altitudes of 1,200-3,000 msl. from Kashmir to Bhutan, and in the Khasi hills in Meghalaya at 1,200-1,500 msl.

Swertia chirata

The plant spreads quickly from seeds which are shed during October-November. The herb can also be cultivated in suitable localities in the temperate Himalayas. The seeds, which are very small, should be shown in a nursery and the seedling transplanted later in the field. The drug (chiretta), obtained from the dried plant, is held in high esteem in India. The whole plant is medicinal, but the root is said to be the most powerful part. The chiretta is much prized in India as a tonic and bitter without aroma or astringency. It enjoys a special reputation in western India as a remedy for bronchial asthma and liver disorders. If taken with sandal wood paste, it is said to stop internal haemorrhage of the stomach. It is anthelmintic, and anti-diarrheal properties in Indian medicinal system. As the species is commonly found in all the northeastern states of India, there is ample scope for production in cultivated form as well as to process them in the states of northeast itself by setting up small scale/cottage units.

Solanum khasianum (Family: Solanaceae):

A stout, much-branched under shrub up to 1.5 m. tall with almost straight prickles, found in Meghalaya, Nagaland, Manipur and Arunachal Pradesh up to an altitude of 1850 m. *S. khasianum* has recently come into prominence as a rich source of solasodine. Fruit of the plant is reported to be a richer source of the steroidal alkaloid than leaves of *S. aviculare*, used in

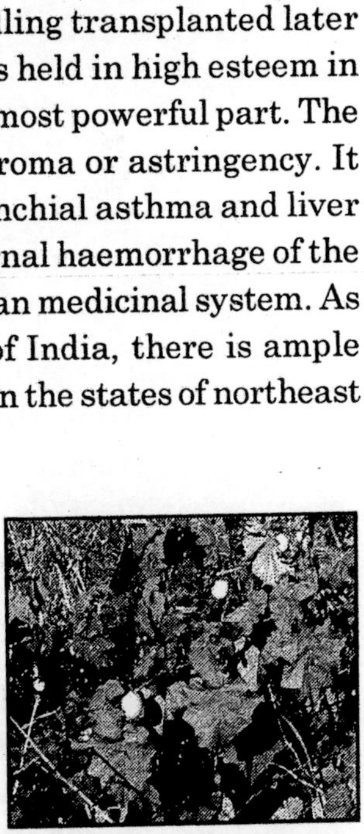

Solanum khasianum

some countries for the production of solasodine. A survey carried out by the Botanical Survey of India has shown that the mature fruit of *S. khasianum* var chatterjeeanum, collected from different areas varies; thus, fruits collected from Cherrapunji (Meghalaya) contain 3.2 per cent; those from Subansiri Frontier Division (N.E.F.A.) and Khasi and Jaintia hills, 2.1 and 2.02 per cent respectively. Fruits of the plants grown in Bombay are reported to yield 2.6 per cent solasodine. The concentration of solasodine is maximum in mature fruits when the colour changes from green to yellow (50-60 days after fruiting). Immature fruits and over-ripe fruits contain very small quantities of alkaloid. The fleshly cover and washed seeds, which form about 60 per cent of the whole fruit, do not contain any alkaloid. The alkaloid is concentrated in the mucilage surrounding the seeds. Leaves contain 0.69 per cent total alkaloids. A new glycoalkaloid, solakhasianin, containing glactose and rhamnose as sugar components has been reported. Fruits contain diosgenin.

Potentilla fulgens **Linn** (Rosaceae) Silverweed: A large genus of herbs or rarely shrubs distributed chiefly in the North Temperate Zone. Nearly 50 species occur wild in India, and some exotics have been introduced into the gardens. *Potentillas* are suitable for borders and rock gardens. They thrive in heavy soils. Propagation is done through seed, cuttings, or division of rootstocks. Potentillas are easily hybridized and fine strains of double blooms have been obtained.

Potentilla fulgens

Potentilla fulgens is an erect, perennial herb, 15-75 cm. high, with a thick rootstock, pinnate leaves and yellow flowers, distributed in the temperate Himalayas from Himachal Pradesh to Sikkim and in the hills of Assam at altitudes of 1,200-4,350 m. The rootstocks are believed to strengthen the gums and teeth and also reported to use in diarrhea.

Potentilla rootstock is dug out from the soil and washed in water. The root bark is peeled off and only the fleshy part is used. Traditional betel chewers used the cut pieces of the fleshy rootstock and chew them fresh with betel nut, betel leaf and lime. The rootstock is bitter (astringent) to taste but when mixed with traditional Khasi betel quid its taste changes favorably. For general health, the peeled rootstocks are boiled in water and taken as such. It is observed that habitual chewers rarely suffer from gum problems and stomach ailments. The watery extract is used as a mouthwash and gargling for treatment of throat infections and infected gums. The rootstocks may be taken for conditions leading to diarrhea, colitis, ulcerative colitis, dysentery and also for piles. The dark brown crystallized mass obtained from methanolic extracts is found to be effective for Dalton's Lymphoma (a type of cancer in mice), antidiabetic, antigastritis, etc.

Piper longum

Piper longum (Piperaceae): A slender aromatic climber with perennial woody roots occurring in the hotter parts of India, from Central Himalayas to Assam, Khasi and Mikir hills, lower hills of Bengal and evergreen forests of Western Ghats from Konkan to Travancore.

The fruits are used as spice and also in pickles and preserves. They have a pungent pepper-like taste and produce salivation and numbness of the mouth. Recent work on the fruit of *P. longum* has shown the presence of the alkaloids piperine (4-5%) and piplartine and two new liquid alkaloids, one of which is designated as alkaloid A. This is closely related to pellitorine producing marked salivation, numbness and tingling sensation of mucous membranes of the mouth. Sesamin ($C_{20}H_{18}O_6$), dihydrostigmasterol and a new sterol, piplasterol are also present. A sample of dried fruit of *P. longum* on steam distillation gave 0.7 per cent of an essential oil with spicy odour resembling that of pepper and ginger oils. Besides fruits, the roots and thicker parts of stem are cut and dried and used as an important drug *(Piplamul)* in the Ayurvedic and Unani systems.

The fruits as well as roots are attributed with numerous medicinal uses, and may be used for diseases of respiratory tract, viz. cough, bronchitis, asthma etc. as counter-irritant and analgesic when applied locally for muscular pains and inflammation; as snuff in coma and drowsiness and internally as carminative; as sedative in insomnia and epilepsy; as general tonic and haematinic; as cholagogue in obstruction of bile duct and gall bladder; and for miscellaneous purposes as antihelmintic, dysentery and leprosy.

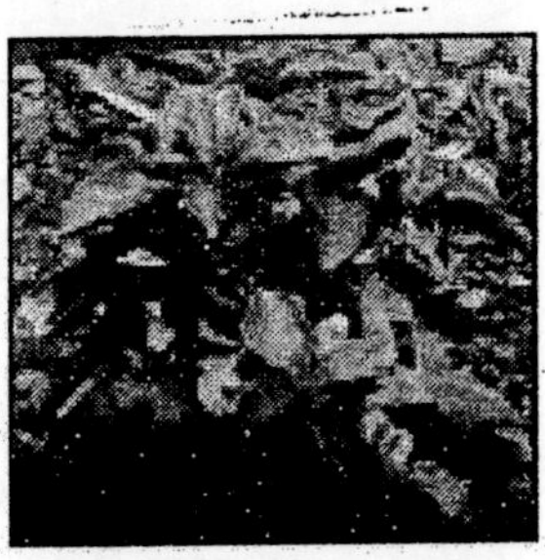

Pogostamon cablin

Pogostemon cablin Benth. Family: Lamiaceae; common name Patchouli. Patchouli is an erect, branched, pubescent herb, 0.5 to 1.0 m high and aromatic when crushed. The leaves are ovate to oblong-ovate, coarse, simple or doubly crenate-serrate, on both surfaces more or less densely tomentose, the glands are dotted beneath, up to 12 cm by 10 cm.

The oil is found mainly in the leaf and a small quantity is present in the tender parts of the stem. The yield of oil varies from 2.5 to 3.5% on shade-dried leaves and an average yield of 2.5% may be considered satisfactory in commercial distillation. The oil of patchouli is viscous liquid, yellowish-green to brownish in colour and possesses a warm, powerful, sweet herbaceous, aromatic and spicy fragrance. It is found that 90% of the oil consists of compounds which have very little influence upon its odour. Sesquiterpene constitutes 40-50% of the oil out of which patchouli camphor or patchouli alcohol, called patchoulol, represents 35-40% of the oil, which is also odourless. However, it is found that one or more satellite components of patchouli may be responsible for the characteristic odour.

Patchouli oil is extensively used as a flovouring ingredient in major food products, including alcoholic and non-alcoholic beverages, frozen dairy desserts, candy, baked goods, gelatine, meat and meat product. It is one of the most important essential oil of the perfumery industry because the oil blends well with other essential oils like sandalwood, geranium, lavender, cedar wood derivatives and clove oils. Hence, it is regarded as the best fixative for heavy perfumes which impart strength, character, alluring notes and lasting qualities. There is no synthetic chemical to replace the oil of patchouli, which further enhances its value and unique position in the perfumery market, and there is great demand for it in soaps, scents, and body lotions, pre-shave and after shave lotions, detergents, tobacco and incense manufacturing factories. The oil also gives one of the finest attars when blended with sandalwood oil. The oil of patchouli is used as an ingredient in insect repellent preparations and is said to have anti-bacterial properties as well.

The fresh leaves of patchouli have medicinal value and are used as a decoction with other drugs to treat nausea, diarrhoea, cold and headache in a Chinese system of medicine. The dried leaves are used for scenting wardrobes, and known for their antirheumatic action. It is also used as a masking agent for alcoholic breath. Tenacity of odour is one of the virtues of patchouli oil and is one of the reasons for its versatile use.

The shade-dried patchouli leaves are subjected to steam-distillation for extracting the oil of patchouli. The dried herb can be immediately distilled or could be store for sometime, according to convenience. But if it is distilled for too long, the oil will have a disagreeable odour.

Taxus baccata (Family: Taxaceae) found in the high altitudes of Sub Himalayan forest, particularly Arunachal Pradesh and Meghalaya of Northeast India. It has a restricted distribution in Himalayas occurring sporadically under mixed coniferous forests at altitude 2,500-2,700 msl. The plant is in great demand nationally and internationally for its leaf and barks as pharmaceutical raw material and has gained importance recently. Local herbal practitioners in West Kameng, Arunachal Pradesh, observed that the price eventually paid to the villager is around Rs. 4/kg of *Taxus* leaves,

Taxus baccata

whereas in Tezpur it is priced at Rs. 35/kg. The illegal collection of *Taxus* ranges from 5 to 6 trucks per trader. *Taxus* yield the drug taxol from its leaves and barks. Taxol is used for the treatment of ovarian and breast cancer. The market price of taxol is far more fantastic, because of its rareness and effectiveness. As per the market price of 1994, one kilogram of taxol costs Rs.180 crores. However, to produce one kilogram of taxol about 2000 to 3000 fully grown *Taxus* trees will be necessary. Therefore it is apparent that the value of about 25,000 trees will far exceed the value of entire forest wealth of a state like Arunachal Pradesh or Meghalaya. In Meghalaya this species was abundant only in the "Law Lyngdoh" sacred grove of Mawphlang. Destruction of its natural location and illegal trade to be checked to conserved it *in situ* as far as possible or grown *ex situ* in protected areas under similar conditions of habitat and climate. Conservation in Botanical and Experimental gardens should be considered, propagation through seed and other technique be attempted.

Smilax glabra (Family: Liliaceae)

The rhizomes of several species of this genus furnish the drug Sarsaparilla. Since botanical description of the various species generally deal with the aerial parts of the plants, details regarding the rhizomes are lacking for most species. *Smilax glabra, Smilax lanceolata, Smilax perfoliata* are some of the Indian species having tuberous rhizomes. *Smilax china* Linn. a native of China, has such tuberous roots which are well known as China root and are employed medicinally in India under the name *Chob-chini*. The rhizomes of many other tuberous species of *Smilax* found in India are used as medicine for the same purposes. The tubers are fleshy when immature but become woody later on; the tubers and young asparagus-like shoots of a number of species are eaten raw or cooked as vegetable. Alcoholic extract of the plant is reported to contain a glycoside and a colouring matter. Decoction of the fresh root is used for the treatment of sores and venereal complaints.

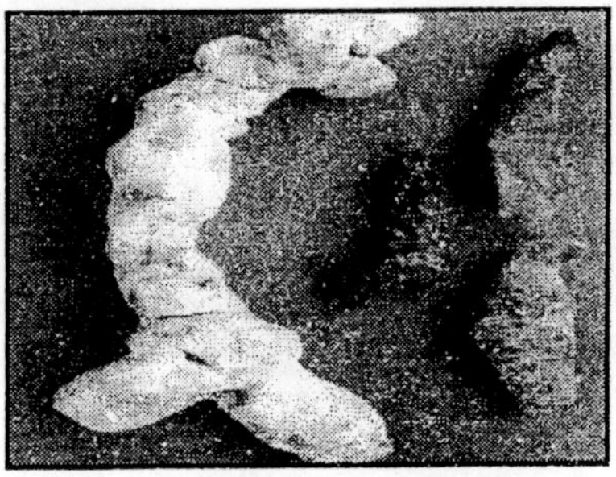

Smilax glabra

Flemingia vestita (Family: Fabaceae)

A much-branched trailing herb with tuberous roots, found practically throughout the Himalayas and Khasi Hills up to an elevation of 8,000 ft. Stems 1-3 ft. long, leaves trifoliolate with obovate-cuneate leaflets; flowers bright red; pods sub-cylindrical, hairly, 1-seeded.

Flemingia vestita

The plant is cultivated in Khasi Hills, often following a potato crop, for its tuberous roots. It is sown in February-May and harvested in August-November. About 4 mounds of 'seed' in the form of tubers are sown per acre and a yield of 30-40 mounds per acre was obtained. The tubers are nearly elliptical, 1.5-2 in. long and possess an agreeable nutty flavour. The skin which is somewhat pungent is removed by rubbing under water or by peeling and the tubers are eaten raw. It has antihelminthic properties.

Curcuma longa syn. ***Curcuma domestica*** (Zingiberaceae) Turmeric's bright yellow colour and spicy taste are familiar to lovers of Indian food, its medicinal value is not so well known. During the last two decades, turmeric's ancient use as a treatment for digestive and liver problems has been largely confirmed by scientific research. The herb has also been shown to inhibit blood-clotting, relieve inflammatory conditions and help lower cholesterol levels. Turmeric is native to India and southern Asia and is cultivated throughout southern and eastern Asia. It is propagated by cuttings from the root, it needs well-drained soil and a humid climate. The rhizome is unearthed in winter.

Curcuma longa

Despite its longstanding use in India and China, the therapeutic action of turmeric were not researched until recent decades when there was an upsurge of interest in food and medicines that lower cholesterol levels or have antioxidant properties (neutralize harmful free radicals) Research since the early 1970s, mainly in India, has confirmed turmeric's traditional actions and revealed potential new uses for it. When applied to the skin and exposed to sunlight, turmeric is a strong antibiotic. Curcumin is the constituent responsible for this action. Curcumin is also a strong antioxidant than Vitamin E. Turmeric may be a valuable preventive remedy for those at risk of developing cancer, but more research is needed. Research has shown that turmeric has an anticoagulant action, keeping the blood thin. It also increases bile production, and has a protective action on the stomach and liver.

Turmeric improves the action of the liver and is a traditional remedy for jaundice in both Ayurvedic and Chinese herbal medicine. It is also an ancient herb for digestive problems such gastritis and acidity, helping to increase mucus production and protection and protects the stomach. The herb also alleviates nausea. Even though turmeric does not relieve pain, its anti-inflammatory action makes it useful for arthritis and other inflammatory conditions such as asthma and eczema. Due to its anti-inflammatory, blood-thinning and cholesterol lowering properties, turmeric is now used to reduce the risk of strokes and heart attack. Applied to the skin, turmeric is useful in treating a number of conditions, including psoriasis and fungal infections such as athlete's foot.

North Eastern Biodiversity Research Cell (NEBRC), Shillong, on its previous reports showed that curcumin content of some varieties of *C. longa* may be as high as 3.0 to 3.6% but a variety of this plant cultivated in Lakadong village and its adjoining areas like Shangpung etc. of Jaintia Hills, Meghalaya, gave a yield higher than what had been reported. The yield of curcumin from this local product was 4.09 and 3.89% of the dried tissue. Identification and quantification of the curcumin extracted was performed in the High Performance Liquid Chromatoghaphy (HPLC) using standard Curcumin (HIMEDIA) as reference standard. The methodology of curcumin extraction employed by the NEBRC may now be used for Value Addition and other commercial purposes.

Table 1
Medicinal plants used by the Khasis of Meghalaya

Sl. No	Scientific Name	Local Name	Family	Status	Mode of Utilisation
1.	*Acorus calamus*	U-bet	Araceae		Leaves boiled in water for 2-3 hours and the vapour inhaled for influenza and headache.
2.	*Adiantum phillipense*	Tyrkhang khyllai	Adiantaceae	R	Paste are applied to fractured bones
3.	*Ageratum conyzoides*	Ksang dngiem	Asteraceae	A	Paste of leaf and lime is applied to cuts which acts as homeostatic.
4.	*Ajuga bracteosa*	Tiew khmut tuta	Lamiaceae	F	Crushed leaves are used as astringent to stop bleeding. Leaf decoction with honey and ginger juice is used for high fever and respiratory congestion.
5.	*Albizzia chinensis*	Dieng phallut	Mimosaceae	C	Bark decoction is applied on ringworm and also as antidote to insect bite
6.	*Allium tuberosum*	Jyllang	Liliaceae	A	Extract of whole plant is used against problems, specially for hypertension

(Contd...)

Sl. No	Scientific Name	Local Name	Family	Status	Mode of Utilisation
7.	*Alysicarpus monilifer*		Fabaceae	F	Whole plant is made to a paste with ginger and mustard oil and used as antidote for snake bite.
8.	*Amomum aromaticum*	Ilashi saw	Zingiberaceae	A	During nausea and vomiting, the rhizome is smashed and made into paste in hot water and taken directly.
9.	*Anaphalis adnata*	Shkor blang	Asteraceae	C	Paste made of leaves and lime applied to cure moematomia.
10.	*Antidesma thwaitesianum*	Soh-syllai	Euphorbiaceae	A	Patients suffering from pains in the joints are bathed with the solution from the boiled leaves.
11.	*Astilbe rivularis*	Pdah	Saxifragaceae	F	The leaves are eaten raw to cure toothache; also given for blood purification
12.	*Averrhoa carambola*	Sohpyrshong	Averrhoaceae	A	Ripe fruit are taken as medicine for jaundice.
13.	*Azadirachta indica*		Meliaceaae	F	Boiled leave extracts are used for diarrhoea and dysentery.
14.	*Bauhinia variegate*	Dieng tharlong	Caesalpiniaceae	C	Flowers are boiled and eaten for piles for dysentery for 6-7 days
15.	*Begonia josephi*	Jajew	Begoniaceae	C	The bulbs are eaten raw in case of stomach pain and indigestion.
16.	*Begonia palmata*	jajewmaw	Begoniaceae	C	Paste made of underground parts applied on swellings due to vaccination. Stems and leaves eaten raw as an antidote for poisoning and vomiting. Stem and leaves

(Contd...)

Sl. No	Scientific Name	Local Name	Family	Status	Mode of Utilisation
					eaten raw for stomach troubles.
17.	*Berberis wallichiana*	Dieng niangmat	Berberidaceae	F	Decoction of young twigs mixed with leaf juice of *Oxalis richardiana* is given for dysentery, diluted decoction of leaves is used for conjunctivitis
18.	*Betula alnoides*	Dienglieng	Betulaceae	C	Root extract is given for indigestion and flatulence
19.	*Biophytum sensitivum*		Oxaladiaceae	F	Leaves paste is applied on the forehead of the patient suffering from headache, giddiness and fever.
20.	*Bonnaya reptans*	Kra-thang syndat	Scrophula-riaceae	F	Juice of crushed leaves mixed with milk and taken to cure urinary ailments.
21.	*Brugmansia suareolens*	Sla-toh-toh	Solanaceae	F	Leaves dried on fire mixed with leaves of *Solanum torvum* and tubers of *Flemingia* sp. ground to a paste is rubbed on body in case of bodyache and also applied on boils.
22.	*Cannabis sativa*	Kynja	Cannabiaceae	A	The leaves and fruits are applied for skin diseases and stomach disorder.
23.	*Capsicum annum*	Soh-mynken syiar	Solanaceae	C	Fruits mixed with leaves of *Dendrobium bakeril, ficus sp., Grewia disperma., Neyraudia meyraudiana, Tinospora cordifolia* and tubers of *Zingiber* sp, and a Vitaceae member *kymbatlamut* (Khasi) are ground to a paste and applied on snake bite.

(Contd...)

Sl. No	Scientific Name	Local Name	Family	Status	Mode of Utilisation
24.	*Cinnamomum glandiferum*	Diengsing	Lauraceae	C	Decoction of leaves and inner bark is given for fever, cold and cough. Fruit paste is applied on rheumatic joints
25.	*Cinnamomum pauciflorum*	Dieng tarthia	Lauraceae	C	Extract of bark and young shoots mixed with coconut oil, is used as an antiseptic
26.	*Cinnamomum tamala*	La tyrppad	Lauraceae	C	Leaves fried in mustard oil and placed on tooth to remove toothache.
27.	*Citrus latipes*	Sohkymphor	Rutaceae	C	Fruit juice is taken as an appetizer; crushed leaves are applied on gouty and rheumatism joints. Juicy slices of fruit are rubbed on rashes and ringworm
28.	*Colocasia esculcenta*	La wang	Araceae	F	Tender leaves tied on forehead in case of high temperature due to fever. Partially cooked corms for rickets disease.
29.	*Conyza bonariensis*	Kynbat symbai pum pum	Asteraceae	F	Decoction and residue of leaves are used as astringents on cuts.
30.	*Costus speciosus*	Sla pangmat	Zingiberaceae	F	Rhizome cut and ground into pieces and the powdered eaten against bronchitis., inflammation & anemia and rheumatism
31.	*Curcuma angustifolium*	Khniang-soh-pet	Zingiberaceaae	F	Paste of rhizome and leaves boiled in water and given to children for gripe.
32.	*Curcuma domestica*	Shynrai stem	Zingiberaceae	A	Rhizome paste with leaves of *Limdera latifolia* and fruits of *Piper longum* applied for various skin

(Contd...)

Sl. No	Scientific Name	Local Name	Family	Status	Mode of Utilisation
					disease; paste of rhizome of this plant, ginger and mustard oil applied and tied around fractures to set the bone.
33.	*Daphne bholua*		Thymelaeaceae	F	Roots are used for intestinal troubles
34.	*Delphinium altissimum*	Bad soh-plihrit	Ranunculaceae	C	Leaves are made into paste and is applied as plaster to glandular swellings
35.	*Dendrobium mochatum*	Tiew dieng	Orchidaceae	F	The leaves juice is used as the ear-drops for ear pain
36.	*Desmodium gangeticum*		Papilionaceae	F	The roots crushed and mixed with ginger are administered for dysentery
37.	*Dirchrocephala bicolor*	Liang poh tiew	Asteraceae	C	Paste make of young leaves applied to wounds and cuts to stop bleeding and quick healing
38.	*Drymaria cordata*	Bat-nongrim	Caryophy-llaceae	A	The whole plant is crushed and the juice is applied for burns, skin diseases and snake bites
39.	*Elephantopus scaber*	Kynbat skrut sriang	Asteraceae	C	Aqueous extract consumed orally to induce abortion, also to treat urinary disorders also used as contraceptive.
40.	*Eleusine indica*	Lang krai	Poaceae	C	Root juice is used for jaundice
41.	*Elsholtziz blanda*	Bat-skain	Lamiaceae	C	The juice of the leaves is applied for mosquito bites and as mosquito repellant
42.	*Engelhardtia spicata*	Dieng Iyba	Juglandaceae	C	Fine paste made from inflorescence and young

(Contd...)

Sl. No	Scientific Name	Local Name	Family	Status	Mode of Utilisation
					leaves is applied on scabies and other skin diseases.
43.	*Erigeron karvinskianus*	Kynbat tiew star	Asteraceae	C	Crushed leaves are applied on cuts or wounds as an astringent
44.	*Erythrina arborescens*	Dieng-song	Papilionaceae	F	The leaves are made into paste with ginger and applied for skin diseases of pig.
45.	*Eupatorium adenophorum*	Bat iong	Asteraceae	A	Crushed leaves applied on injuries
46.	*Fiscus virens*	Dieng soh phoh khlaw	Moraceae	C	Leaves are boiled and given for loss of appetite
47.	*Garcinia cowa*	Soh syrum	Clusiaceae	F	Extract of crushed fruits mixed with little lime added to boiled water and the vapour inhaled for severe headache and cold The fruit is powdered after sun drying and used for dysentery
48.	*Garuga pinnata*	Dieng khiang	Burseraceae	F	Fruits juice is given for indigestion, stem juice is applied for conjunctivitis and leaf juice with honey is given for asthma
49.	*Glochidion khasicum*	Ja lwai	Euphorbiaceae	F	Leaves eaten for dysentery and associated stomach troubles
50.	*Gmelina arborea*	Dieng laphiang	Verbenaceae	F	Berries are used as a purgative. Root juice is used as antidote for snake bite and insect stings
51.	*Gomphostemma parviflora*		Labiatae	F	Leaves paste is applied on the forehead of the patient suffering from headache, giddiness and fever.

Figure 1. Few Available photographs of selected species of medicinal plants

Acorus calamus

Adiantum phillipense

Averrhoa carambola

Begonia sp

Bauhinia sp

Cinamomum tamala

Citrus latipes

Costus speciosus

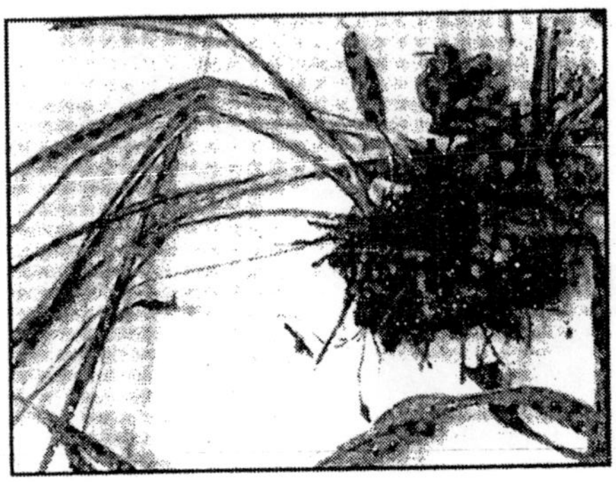

Curcuma sp

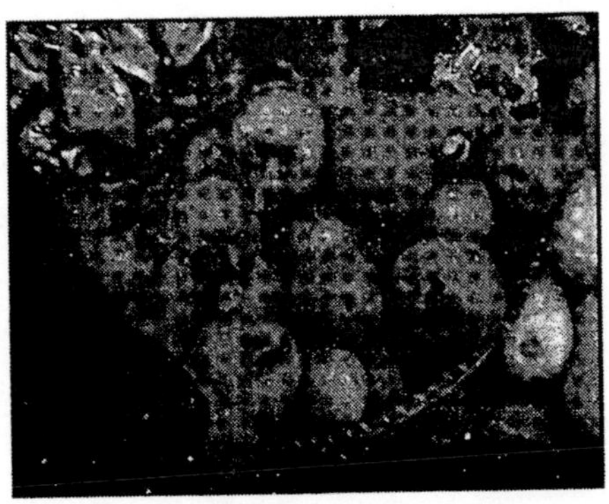

Garcinia Sp

4. Separation Methods Database

Powdered Leaves / Leaves & Inflorescences of Medicinal Plants

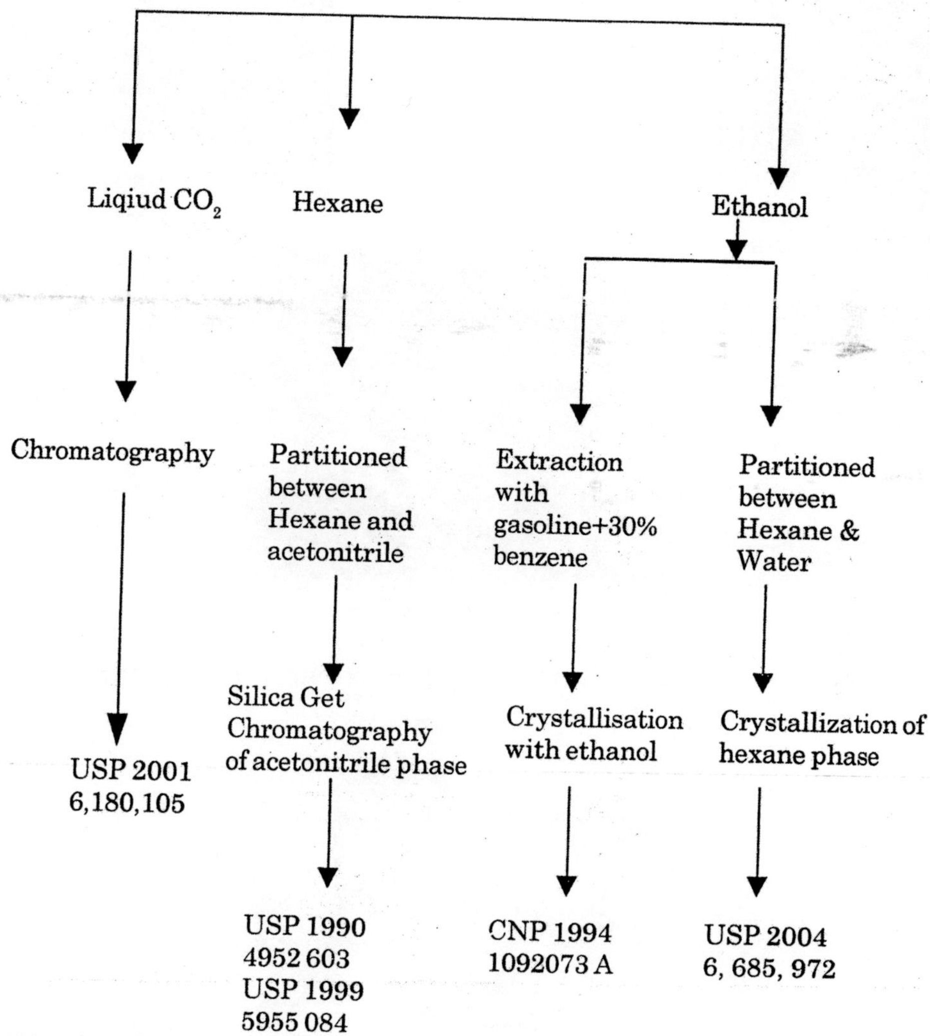

Figure 2. Sushil Kumar & Sachi Srivastav. Establishment of Artemisia combination therapy as first line of treatment for combating malaria: *Artemisia annua* cultivation in India needed for providing sustainable supply chain of Artemisinin. *Current Science*, Vol. 89, No. 7 (2005) pp. 1097 – 1102

Figure 3. Separation Methods by Thin Layer Chromatography (TLC)

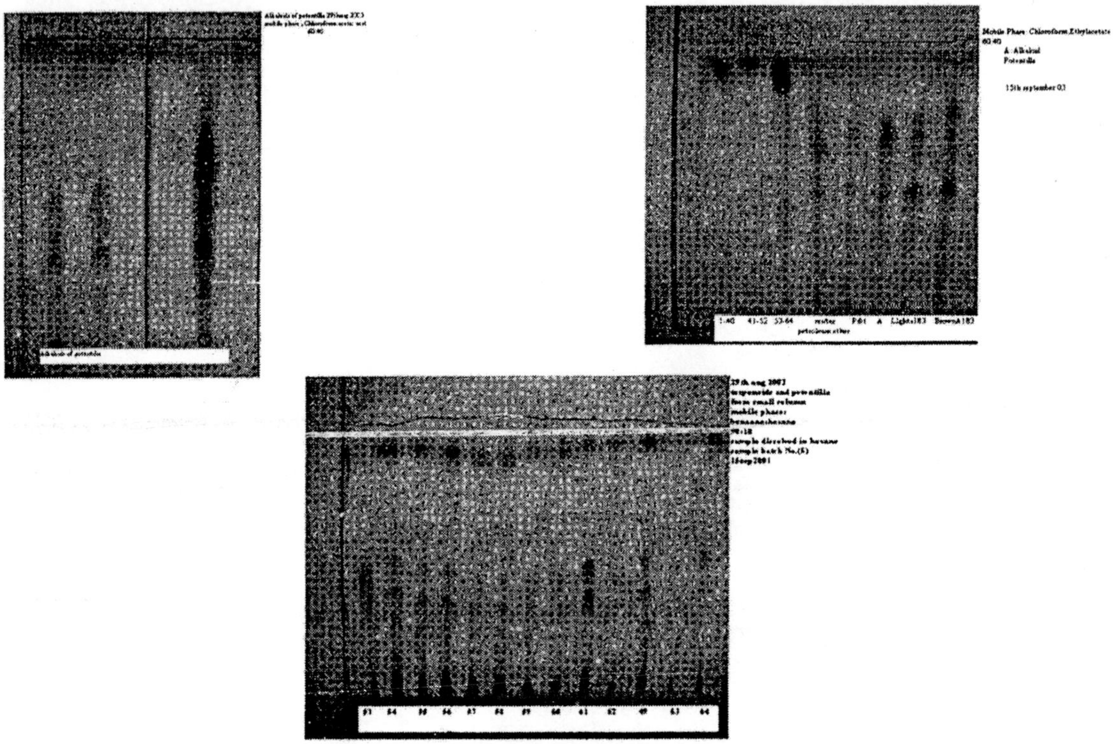

Figure 4. Steam Distillation Unit for Extraction of Essential Oils

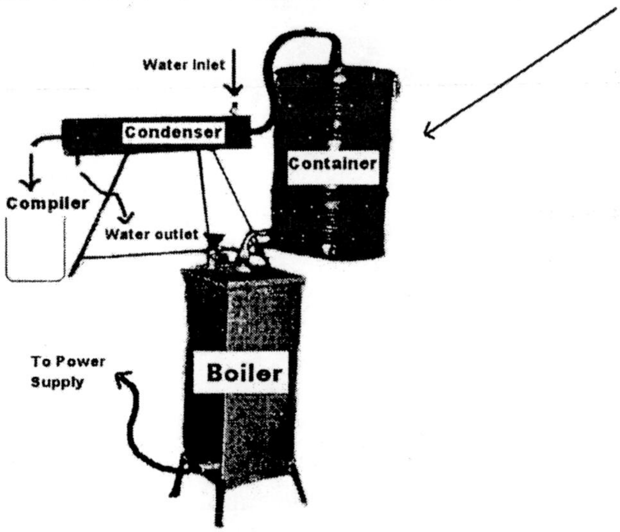

Designed by NEBRC, NEHU

5. Commercial Database of Medicinal Plants

Collectors, Collection and Storage

Medicinal plant collectors are usually poor villagers. Plant collection is their part time activity besides agriculture, farming and live stock keeping. The collectors include 40% womenfolk, 35.0% men folk and 25% children. They collect medicinal plants during spring and summer seasons which starts from April to September and sell it in the local market to fetch some money. One can see a person coming from a hilltop with a bundle of fuelwood on his head and a bag of medicinal plants in his hands. He handovers his bag of medicinal plants to local Practitioners and put in his pocket whatever money he gets. About 500 families are involved in medicinal plant collection and they collect 5000 tons of medicinal plants annually (Choudhary *et al.* 2000). The availability of medicinal plants has decreased in the last twenty years. However, in the last few years pressure on medicinal plants collection is relaxed due to an increased cultivation.

Bulk of medicinal plants is collected and the collectors sell them in local markets while some of them are kept in homes for curing different ailments. In the figure below a chain of people involved in medicinal plants marketing. Major proportions of plants collected are sold fresh while some plants are stored in bags and sacks from one week to one year. Before storing, these plants are washed and kept under the sun for drying. During storage considerable amounts of medicinal plants are wasted due to humidity, insect attacks, inappropriate storage facilities and lack of awareness on the part of collectors.

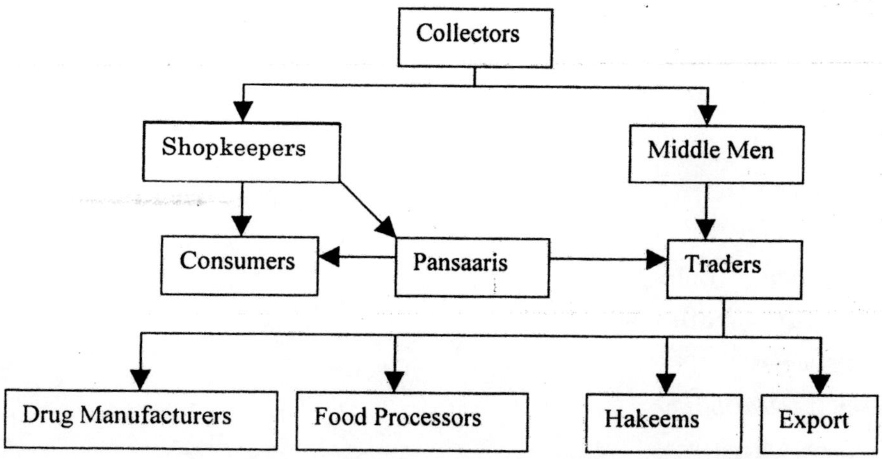

Figure 5. Chain of people involved in the medicinal plant trade in Northeast India

Table 2
Export of the following Medicinal plants would be permitted under legal procurement certificate (L. P. C.)

Sl. No	Scientific Name
1	*Acontium species*
2	*Acorus species*
3	*Adhatoda beddomei* C. B. Clarke
4	*Ampelocissius indica*
5	*Angelica glauca* Edgew
6	*Angiopteris species*
7	*Aquilaria malaccensis* Lam
8	*Aristolochia species*
9	*Arnebia bethamili* (Wall. Ex. G. Don)
10	*Artemisia species*
11	*Arundinaria jaunsarensis Gramble*
12	*Atropa species*
13	*Balanophora species*
14	*Beddomes cycad*
15.	*Berberis aristata* DC.
16.	*Berberis Kashmirana*
17.	*Berberis kashmirana*
18.	*Berberis petiolaris*
19.	*Berberis lycium* Royle
20.	*Berberis atistata* DC
21.	*Blue Vanda*
22.	*Munium persicum* (Boiss) Fedts
23.	*Cayratia pedata* (Wall) Gagnep
24.	*Ceropegia species*
25.	*Colchicum luteum baker*
26.	*Commiphora whiggtii*

(Contd...)

Sl. No	Scientific Name
27.	*Coptis species*
28.	*Coptis teeta* Wall
29.	*Coscinium fenestratum*
30.	*Costus speciousus* (Koenig) Sm.
31.	*Craterostigma plantagineum*
32.	*Curcuma caesia* Roxb
33.	*Cyatheaceae species*
34.	*Cycadaceae species*
35.	*Cyclea fissicalyx*
36.	*Decalepis hamiltonii* Wight & Arn
37.	*Didymocarpus pedicellata* R. Br.
38.	*Dioscorea deltoides* Wall
39.	*Dolomiaea pedicellata*
40.	*Drosera species*
37.	*Dipterocarpus indicus* Bedd
38.	*Dysoxylum malabaricum* Bedd
41.	*Ephedra species*
42.	*Eulophia cullenii*
43.	*Eulphia ramemtacea*
44.	*Euphorbua species*
45.	*Frerea indica*
46.	*Fritillaria roylei*
47.	*Garcinia travancorica* Bedd
48.	*Gentiana kurroo* Royle
49.	*Glorcinia travancorica*
50.	*Gloriosa superba* Linn
51.	*Gnetum species*
52.	*Gymnema khandalense*
53.	*Gymnema montanum* Hook. F.
54.	*Gynocardia odorata* R. Br.

(Contd...)

Sl. No	Scientific Name
55.	*Hedychium coronarium* Koenig
56.	*Hedychium spicatum* Buch-Ham
57.	*Hellotrophium keralense*
58.	*Humboldtia vabliana*
59.	*Hydrocarpus alpina* Wight
60.	*Hydrocarpus species*
61.	*Hyoscyamus niger* Linn
62.	*Ilex khasiana*
63.	*Inula racemosa* Hook. F.
64.	*Iphigenia indica* Kunth
65.	*Janakia arayalphathra*
66.	*Kaempferia galanga* Linn
67.	*Kingiodendron pinnatum* (Roxb.) Harms
68.	*Ladies slipper orchids*
69.	*Lamprachatenium microcepphalum benth*
70.	*Luvunga scandens* (Roxb.) Buch. Ham.
71.	*Madhuca diplostemon*
72.	*Madhuca longifolia* (Koenig)
73.	*Meconopsis aculeata* Royle
74.	*Meconopsis betonicifolia*
75.	*Nardostachys species*
76.	*Nervilla aragoana Gaudich*
77.	*Nilgirianthus ciliatus* (Nees) Bremeck
78.	*Orchidaceae species*
79.	*Osmunda species*
80.	*Panax pseudoginseng*
81.	*Physochlaina pratealta* (G. Don) Miers
82.	*Picrorhiza kurroa* Royle ex benth
83.	*Piper barberi* Gamble

(Contd...)

Sl. No	Scientific Name
84.	*Pitcher Plant*
85.	*Padphyllum hexandrum* Royle
86.	*Praltia serpumlia*
87.	*Prezwalskia tangutica*
88.	*Pterocarpus santalinus* Linn. F.
89.	*Rauvolfia serpentina* Benth ex Kurz
90.	*Red Vanda*
91.	*Rheum emodi* Wall ex Meissn
92.	*Rheum nobile* Hook.f. & Thoms
93.	*Rhododendron species*
94.	*Salacia oblonga* Wall ex Wright & Arn
95.	*Salaca reticullata* Wright
96.	*Saussurea gossypiphora* D. Don
97.	*Saussurea lappa* C. B. Clarke
98.	*Saussurea obvallata* Wall ex C. B. Clarke
99.	*Saussurea simpsoniana*
100.	*Shorea tumbuggaia* Roxb
101.	*Swertia chirayita* (Roxb. Ex. Flem) Karst
102.	*Strychnos aenea*
103.	*Strychnos Potatorum* Linn
104.	*Swertia lawii Burkill*
105.	*Syzgium travancoricum*
106.	*Taxus wallichiana*
107.	*Trichopus seylanicus*
108.	*Trichosanthes anamalaiensis*
109.	*Urginea species*
110.	*Utleria salicifolia*
111.	*Vateria macrocarpa* Gupta
112.	*Valeriana jatamansi* Jones
113.	*Valeria lechenauli* DC

Table 3
Medicinal plants exported as raw drugs

Sl. No.	Trade Name	English Name	Botanical Name
1.	Agar	Eagle wood	*Aquilaria agallocha Roxb*
2.	Ajmod	-	*Apium leptophyllum*
3.	Amba Haldi	Mango Cinger	*Curcuma amala*
4.	Arjun	Arjun	*Terminalisa arjuna*
5.	Arnimool	-	*Clerodendrum phlomidis*
6.	Ashok	Ashoka	*Saraca asoca*
7.	Ban Tulsi	-	*Majorana hortensis*
8.	Bhringraj	-	*Eclipta alba*
9.	Brahmi	Thyme Leaves Gratiola	*Bacopa monieri*
10.	Chawal Mongra	-	*Gynocardia odorata*
11.	Chiraita	Indian Gentian	*Swertia chirata*
12.	Chobcchini	China Root	*Smilax china*
13.	Daruhalsi	Indian Berberry	*Berberies aristata*
14.	Gaj Pippali	-	*Scindapsus offinalis*
15.	Gilo	Moon Creeper	*Tinospora cordifolia*
16.	Gudhal	China Rose	*Hibiscus rosa-sinensis*
17.	Inderjau	Shoe flower	*Holarrhena pubescens*
18.	Jivanti	-	*Leptadenia reticulata*
19.	Kachur	Zedoary	*Curcuma zedoaria*
20.	Kalonji	Small Fennel	*Nigella sativa*
21.	Kapoor Kachri	-	*(i) Kaempferia galanga* *(ii) Hedychium spicatum*
22.	Kateli	Febrifuge Plant	*Solanum Galanga*
23.	Kulanjan	Siamese Ginger	*Alpinia galanga*
24.	Kusum Phool	Safflower	*Carthamus tinctorius*
25.	Kutaj	Ester Tree	*Holarrhena antidysenterica*
26.	Lodh	Symplocos Bark	*Symplocos racemosa*
27.	Majith	Indian Madder	*Rubia cordifolia*
28.	Meda	-	*Polygonatum cirrihifolium*
29.	Mushkdana	Musk Mellow	*Hibiscus abelmoschus*
30.	Nagbeli	-	*Lycopodium clavatum*

(Contd...)

Sl. No.	Trade Name	English Name	Botanical Name
31.	Nageshwar Phool	Iron wood	*Mesua ferea*
32.	Nisoth	Terpeth Root	*Operculina turpethum*
33.	Palsh Papda	Bengal Kino Tree	*Butea monosperma*
34.	Parul	-	*Stereospermum suaveolens*
35.	Pipal (Borki)	Long pepper	*Piper longum*
36.	Pipal (Gol)	-	*Piper mullesua*
37.	Pipal (Savali)	Long Pepper	*Piper longum*
38.	Priyangu (Phool)	-	*Callicarpa macrophylla*
39.	Punarnava	-	*Boerhaavia diffusa*
40.	Ratti	-	*Abrus precatorius*
41.	Rohitak	Rohida Tree	*Tecomella undulata*
42.	Rudraksh		*Elaeocarpus ganitrus*

Table 4
Profit received by collectors and wholesalers on medicinal plants

Name of plant	Collector rate (Rs.)	National rate (Rs.) market	International rate (Rs.) market
Aconitum napellus	120	500	3000
Bergenia himalacia	25	250	1500
Carum carvi	200	400-500	2000
Carthamus tinctorious	150	500	1500
Dactylorrhiza hatagirea	——	500/ 1250	7000
Ephedra gerardiana	5	25	800
Glycyrrhiza glabra	10	30	1200
Hyppopha rhamnoides	45	300	500
Picrorhiza kurrooa	200	——	4500
Podophylum emodi	150	400	500
Sussurea lappa	150	150	2500
Thymus serphylum	35	150	1500
Valeriana walllichii	——	500	5000

Table 5
Medicinal plants on international markets

Botanical Name	Part Sold	Export to
Cichorium intybus	Roots/ seeds	Italy, Hong Kong
Berberis aitchisonii	Roots	Middle East
Bunium persicum	Fruits	Germany, South Korea, South Africa
Juniperus macropoda	Fruits	ASEAN
Onosma hispidum	Roots	Europe
Plantago major	Seeds	Korea, France
Rheum emodi	Roots	Iran, Middle East
Rosa webbenia	Flowers	Korea
Thymus serphyllum	Leaves	Europe, Middle East
Juglans regia	Fruits	Middle East, Europe

6. Recommendations

- Local community should be trained regarding methods of collection and drying, identification and sustainable harvesting.
- Plant collectors should be encouraged and educated in order to enhance their bargaining power, thus increasing their revenues from medicinal plants.
- The locals should be educated about the importance of medicinal plants to their socioeconomic conditions and ecosystem.
- There should be more coordination and cooperation among various agencies such as government departments and pharmaceutical firms pertinent to the utilization and regeneration of selected medicinal plants.
- Traditional medicines manufacturers and scientists should be encouraged to carry out research into the cultivation of medicinal plants.
- Cultivation of threatened medicinal plants should be encouraged in order to relieve pressure on these plants.
- Direct links between collectors and consumers should be established to secure higher level prices.
- Taxonomy of the plants used with local names should be recorded.
- Pictures of the plants or parts used indicating the time of availability is needed.
- Methods of extraction that are used to isolate crude products (TLC, HPLC, etc) must be standardized.

- The IK/TK related to it and what types of diseases it cures must be recorded.
- Scientific experiments on the basis of bioassays done to validate the claims made by IK/TK or healers.
- The profile of DNA, RNA or Proteins that are there in the plant/parts should be done.
- All the above, with whatever photographic documentation which can be obtained, so that the whole information will be put up in a page for easy access. This will help in IPRs, Patents and Claims of ownership.

REFERENCES

Anonymous. 1997. *The Wealth of India*. Vols 1 – 10. Council of Scientific and Industrial Research, New Delhi, India.

Balick Michael J. and Paul Alan Cox. 1996c. Plants that heal. In their *Plants, People, and Culture: the science of ethnobotany*. New York, Scientific American Library, p.25-61.

Chaudhary M. S., Ahmad A.Ali., Sher H. and Malik S. 2000. *Technical Report on Market study of medicinal herbs in Malakand, Peshawar, Lahore and Karachi*. SDC- Intercooperation, Peshawar.

Ghosal S., Saini K.S., Razdan S. and Kumar Y. 1985. Chemical constituents of Amaryllidaceae. Part 12. Crinasiatine, a novel alkaloid from Crinum asiaticum. *Journal of Chemical Research*, Synopses, 3:100-101.

Ghosal S., Shanthy A. and Singh S. K. 1988. Chemical constituents of Amaryllidaceae. Part 27. Isocraugsodine, an N-arylidenephenethylamine from Crinum asiaticum and its E-Z isomerism. *Phytochemistry* 27(6): 1849-52.

Gogoi P. and Ahmed M. 2006. Biodiversity Conservation in Northeast India. *Biodiversity in Northeast India* (ed. B. Kharbuli, D. Syiem and H. Kayang):42 – 46.

Jain S.K. and De Filipps R.A. 1991. *Medicinal Plants of India*, Volume Two. Reference Publications, Inc.

Joseph J. and Kharkongor P. 1997. A preliminary ethnobotanical survey in Khasi and Jaintia Hills, Meghalaya: *Contribution to Indian Ethnobotany* (ed. S.K. Jain). 187 – 194.

Kharduit J. 1999. Medicinal Plants: Utilization and conservation in Meghalaya. *Biodiversity Northeast India Perspective* (ed. B.Kharbuli, D. Syiem and H. Kayang): 89 – 92.

Lalramnghinglova H. 2006. Ethnobotanical Interpretations and future Prospects of Ethnobotany in Northeast India. *Biodiversity in Northeast India* (ed. B.Kharbuli, D. Syiem and H. Kayang):10 – 21.

Michael Buam D.R. and Marngar D. 2006. Medicinal Plants of North Tripura. *Biodiversity in Northeast India* (ed. B.Kharbuli, D. Syiem and H. Kayang): 99 – 105.

Mukerji B. 1953. *The Indian Pharmaceutical Codex.* Vol. I-Indigenous Drugs. Council of Scientific and Industrial Research, New Delhi, India.

Shulz Volker., Rudolf Hänsel and Mark Blumenthal. 2001. Medicinal plants, phytomedicines, and phytotherapy. In their *Rational Phytotherapy: a physicians' guide to herbal medicine.* 4th ed. Berlin, New York, Springer, p.1-39.

Skidmore-Roth Linda. 2004 c. Introduction. In her *Mosby's Handbook of Herbs & Natural Supplements.* 2nd ed. St. Louis, Mosby, p. ix-xxii.

4

Current Status and Future Strategy for Development of Medicinal Plants Sector in India

S. P. Saikia, B. S. Bhau,
Kalyani Medhi and P. B. Kanjilal

1. Medicinal Plants: An Overview

India is located in South Asia between latitudes 6^0 and 38^0N and longitudes 69^0 and 97^0E is rich in biodiversity due to immense variety of climate and altitudinal conditions with varied ecological habitats. The Eastern Himalayas and the Western Ghats represent two hotspots of the biological diversity, which is the reason why India has been identified as one of the 12-megadiversity centers of the world. The other countries being Brazil, Colombia, China, South Africa, Mexico, Venezuela, Indonesia, Ecuador, Peru, USA and Bolivia. Across the country, the forests of India are estimated to harbour 90% of India's medicinal plants diversity in the wide range of forest types that occur. Only about 10% of the known medicinal plants of India are restricted to non-forest habitats. According to Schippmann, *et al.* 2002, India has 20% plant species of medicinal value against the world average of 12.5% (Table 1).

Table 1

Numbers and plants used medicinally worldwide (Schippmann, *et al.* 2002)

Country	Plant species	Medicinal plant species	Percentage
China	26,092	4,941	18.9
India	15,000	3,000	20.0
Indonesia	22,500	1000	4.4
Malaysia	15,500	1,200	7.7

(Contd...)

Country	Plant species	Medicinal plant species	Percentage
Nepal	6,973	700	10.0
Pakistan	4,950	300	6.1
Phillippines	8,931	850	9.5
Sri Lanka	3,314	550	16.6
Thailand	11,625	1,800	15.5
USA	21,641	2,564	11.8
Vietnam	10,500	1,800	17.1
World	422,000	52,885	12.5

But according to Hamilton, 2003, India has about 44% of flora, which is used medicinally (Table 2).

Table 2

Numbers and percentage of medicinal plant species recorded for different countries and regions (Hamilton, 2003)

Country or Region	Total no. of native species of flora	No. of species of medicinal plants	% of flora which is medicinal
China	27,100	11,146	41
India	17,000	7,555	44
Mexico	30,000	2,237	7
North America	20,000	2,572	13
World	297,000-510,000	52,885	10-18

Although it is difficult to estimate the number of medicinal and aromatic plants present worldwide, the fact remains true that India with rich biodiversity ranks first in percent flora, which contains active medicinal ingredients.

2. Medicinal Plants and Traditional Medicine

Both plant species and traditional knowledge are important to the herbal medicine trade and the pharmaceutical industry whereby plants provide raw materials and the traditional knowledge prerequisite information. India has one of the richest plant medical traditions in the world. There are estimated to be around 25,000 effective plant-based formulations, used in folk medicine and known to rural communities in India. There are over 1.5 million practitioners of traditional medicinal system using medicinal plants in preventive, promotional and curative applications. It is estimated that there are over 7800 medicinal drug-manufacturing units in India, which consume about 2000 tonnes of herbs annually. According to an estimate of the World Health Organization (WHO), nearly 80% of the populations of developing countries rely on traditional medicine, mostly plant drugs for their primary health care needs. Traditional medicine has served as a source of alternative medicine, new pharmaceuticals and health care products.

3. Threats of Medicinal Plants in India

Some of the major threats are:

(i) ***Over-exploitation:*** The herbal medicines trade is booming business worldwide. In India there are 46000 licensed pharmacies manufacturing traditional medicines from herbs, 80% of which are from wild. To meet this demand intensive exploitation of medicinal plants diminished the resources of herbs and some species become highly vulnerable.

(ii) ***Unscientific exploitation:*** Plants are being extracted without knowing its mode of reproduction, time of seeding; hence re-growth, replacement and the realistic substitution are hampered.

(iii) ***Environmental degradation:*** Natural calamities like floods, cyclones, typhoons, shifting sand dunes, earthquake and various types of soil erosions lead directly to the demise of species and habitats.

(iv) ***Human population:*** Exploding human populations are degrading the environment at an alarming rate, particularly in tropics. The population pressure has resulted in deforestation which leads to ozone depletion, greenhouse gas, industrial pollution and acid rain which ultimately effect on plant species. Indiscriminate use of pesticides is another threat to biological diversity

(v) ***Intensive cultivation:*** Expansion of cash crops and intensified shifting cultivation has rapidly squeezed out native plant species of many medicinal plants.

(vi) ***Genetically modified organisms/crops:*** According to National Biodiversity Strategy and Action Plan, the biggest and recent threat is the large-scale introduction of exotic and GM crops. These may effects on plant biodiversity by interfering in the life cycle of plant like disease resistance, flowering and fruiting behaviour. Although this is a very controversial issue.

4. Measures for Conservation

A. *In-situ* Conservation: Important *in-situ* conservation methods are:

i. ***Establishment of gene banks***: Areas of greatest genetic diversity should be demarcated and protected from human interference so that the plants can continue to evolve. At present there are 87 National Parks and 447 wildlife sanctuaries extending over an area of about 1.5 lakh Sq. Km which is more than 4.5% of the geographical area of the country.

ii. ***Preservation of natural biodiversity in Biosphere Reserve:*** There are areas of high endemism and biological diversity that posses rich genetic wealth of wild relative of crop plants. These are located in Humid Tropical Regions of Western Ghats, the hilly tracts of Northeast India and in the Temperate Himalayas.

iii. ***Controlled exploitation of naturally occurring medicinal plants:*** Many tribes collect plants for their use in different ailments without destroying mother plant and thus the species are safe from extinction. Same method may be followed by the Government organization, Co-operative and traders.

iv. ***Adopting extension and educational programme:*** An inventory may be prepared for all the medicinal plants grown in natural habitats and distributing this information to village people along with necessary training related to identification, conservation and sustainable use. This will generate awareness among the people about the importance as well as conservation of the species.

v. ***Sacred groves:*** Sacred groves are the old concepts for protection of animals and plants. Tribal people keep some pocket of forest undisturbed and thereby the important vegetation resources protected representing the original flora and fauna of the locality. The concept of sacred groves stops deforestation and leads to the formation of Biosphere Reserve. Some Regional Centres have taken up the study of sacred groves under the forest protection and documentation.

B. *Ex-situ* Conservation

Ex-situ conservation refers to the conservation of germplasm away from its natural habitat.

i. **Botanical gardens:** Botanical gardens can play a key role in *ex-situ* conservation of plants, especially those facing threat of extinction.

   In India important Botanical gardens are:

   a. Indian Botanical Garden, Kolkata

   b. National Botanical Garden, NBRI, Lucknow

c. Lloyd Botanical Garden, Darjeeling
d. Botanical Garden, FRI, Dehradun
e. Tropical Botanic Garden, Mysore

ii. ***Field gene bank of medicinal plants:*** The scope of Field gene bank is to preserve a broad spectrum of genetically diverse medicinal plants.

3 gene banks for medicinal and aromatic plants have been established:

a. National Bureau of Plant Genetic Resources, New Delhi
b. Tropical Botanical Garden & Research Institute, Thiruvananthapuram
c. Central Institute for Medicinal & Aromatic Plants, Lucknow

During the past few years more than 30,000 accessions of 250 medicinal and aromatic plants of tropical region of India are conserved.

iii. ***Seed/Propagule banking system for Medicinal plants:*** Seed and propagules including corm, bulb, rhizome, tuber cutting, synthetic seed are the basic requirement of plant propagation and also for biodiversity conservation and development.

In India, the following organizations are involved in seed-propagule banking systems of plant biodiversity:

a. IPGRI - International Plant Genetic Resources Institute, Delhi
b. NBPGR - National Bureau of Plant Genetic Resources, New Delhi
c. ICFRE - Indian Council of Forestry Research & Education, Dehradun
d. ISST - Indian Society of Seed Technology. New Delhi
e. SAI - Seed Association of India, New Delhi
f. ICRISAT - International Center for Research in Semi-arid Tropics, - Hyderabad
g. CAZRI - Central Arid Zone Research Institute, Jodhpur

Advantages of Cultivation of Wild Medicinal Plants

1. Conservation of water
   - The wild plants are able to tolerate the unavailability of continuous exogenous water supply for reasonable periods.
2. Sustainable development
   - Do not introduce new weeds or other new pests to the ecosystem.
   - Do not need the use of pesticides.
   - Minimal ecological consequences for the agro-ecosystem.

3. Improvement of economy of wasteland

4. Environmental protection
   - Wildlife habitat
   - Biodiversity conservation

5. Economical improvement
   - Fill a gap in the domestic needs such as folk medicine
   - Potential for creation of small industries
   - Provides cash income for settlers in newly reclaimed land.

5. Current Status of Medicinal Plants Chain in India

Most of the medicinal plants are collected from the forests and rangelands. Forest Department of every State is responsible for regulating the collection of species from the wild that are not considered endangered. It determines the areas from which plants can be collected, fixes the volumes to be collected and monitors collection activities in order to prevent illegal and excessive collection. The respective State Forest Department issues permits to medicinal-plant cooperatives of that particular State, which in turn employ contractors to organize collection. The contractors employ local contractors usually farmers with small land holdings who in turn employs collectors who are landless labourers. The contractors sell the collected materials either to the cooperatives or directly to independent traders after paying royalties to the cooperative. The cooperatives sell to either the local agents of wholesalers, or traders or drug manufacturers. The traders supply the domestic and international markets.

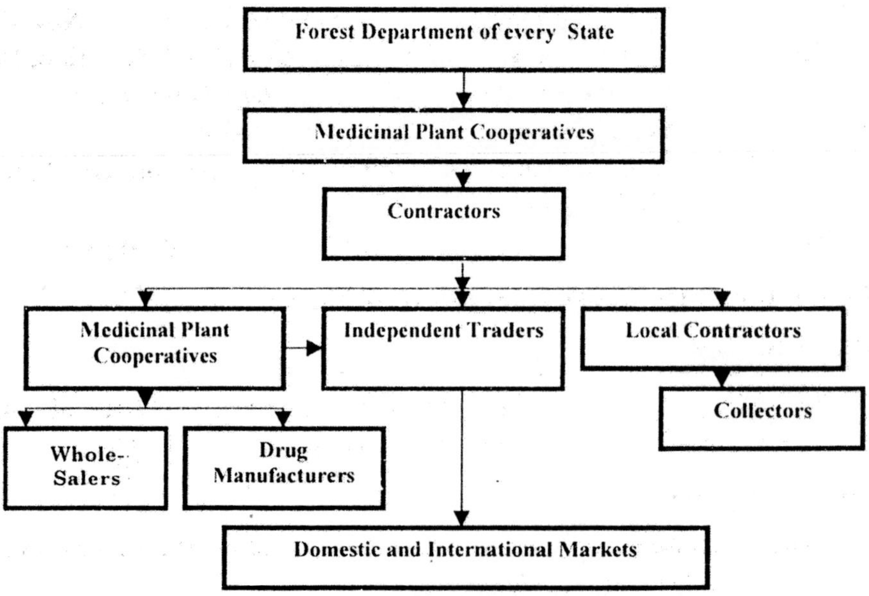

Figure 1. Medicinal-Plant Chain in India

6. Factors Limiting Medicinal Plant Cultivation

The main factors that prevent small holders from taking up cultivation of medicinal plants are:

1. Long gestation period

(i) Many medicinal plants can be harvested only after three years or more. This is particularly true for the plants grown in high altitude areas.

(ii) As most farmers are poor, have small land-holdings and lack credit, they cannot wait so long for returns. Understandably, they are reluctant to convert a significant part of their land to medicinal-plant production.

2. High risk

(i) The cultivation of medicinal plants is also highly risky.

(ii) It is a comparatively new activity and reliable cultivation technologies and other inputs are yet to be fully developed, also the risk of failure is particularly high.

(iii) The farmers also face serious market-related risks and difficulties.

(iv) In most cases they do not have a guaranteed market and price premiums for cultivated material.

(v) They also lack reliable market information about demand and pricing, which puts them in a vulnerable position.

3. Transaction costs

In most of the rural areas of India, the physical infrastructure is poor: road networks are not well developed, poor communication networks limit access to information, and agro-processing facilities are limited. For these reasons the transaction costs for rural producers and local entrepreneurs are high.

4. Social capital and values

As the medicinal-plant trade based on cultivated material is new, various linkages essential for trade are not yet well developed. In the current system the risks related to the level of trust worthiness of the actors involved and the chance that arrangements are not respected are high.

5. Overcoming constraints by public-private partnerships

Public-private collaborations can play an important role in removing many of the bottlenecks. In fact, some promising public-private collaborations can overcome some of the existing impediments to the development of a medicinal-plant chain based on cultivated material. Public-private collaboration will reduce the high transaction costs to acceptable levels, reduce risk, and secure market access to small producers. Thus, such collaborations can overcome many of the constraints and provide a promising mechanism for establishing the conditions necessary for the growth of chains that are in the initial stages of development. There are several private sectors and government Research and Development Institutions in India of which a few have been mentioned.

Among the industry oriented Research and Development Institutes are Dabur Research Foundation, Himalaya Health Care, Zandu Pharmaceuticals, Reliance Life Sciences, Hamdard etc. They carry out research in diverse areas like ayurvedic research which relates traditional knowledge with modern science, pharmaceutical research, phyto-pharmaceuticals, biotechnology, personal care products etc. Apart from these there are several government established Research and Development Institutions like Central Institute of Medicinal & Aromatic Plants (CIMAP), National Botanical Research Institute (NBRI), Central Drugs Research Institute (CDRI), North East Institute of Science & Technology (NEIST) etc. They are committed to provide global standards for plant based research, processing and products to ensure sustained clientele from agriculture, society and industry.

The government of India has also set up the National Medicinal Plant Board (NMPB) to coordinate matters relating to medicinal plants, including developing policies and strategies for conservation, proper harvesting, cost-effective cultivation, Research and Development, processing and marketing of raw materials to protect, sustain and develop the sector. Besides there are several other agencies both in private sectors and under the government sectors that are involved in research and development activities of medicinal plants, along with conservation and upbringing of new plant-based products in national and international markets.

7. Conclusion

A major factor impeding the development of the medicinal plant based industries in India has been the lack of information on the social and economic benefits that could be derived from the industrial utilization of medicinal plants. Except for the use of these plants for local health care needs, not much information has been available on their market potential and trading possibilities. India has the knowledge and skill to develop its R&D capabilities. It is the second largest exporter of medicinal plants. Instead of exporting such a large amount of valuable resource with very low returns it can think about developing its own R&D capabilities and produce finished goods in the form of modern medicines and health care products derived from plant origin and based on the knowledge of alternative system of medicine. India has to take up the challenge of leading the drug and herbal market while conserving its rich heritage through proper planning and implementation of policies.

REFERENCES

Belt J., Lengkeek A. and Vander Zant J. 2003. Cultivating a healthy enterprise: developing a sustainable medicinal plant chain in Uttaranchal – India. KIT Publishers, Amsterdam. *KIT Bulletin* No. 350.

de Silva T. 1997. Industrail utilization of medicinal plants in developing countries. In: Bodeker, G., Bhat, K. K. S., Burley, J. and Vantomme, P. eds. *Medicinal Plants for Forest Conservation and Health Care.* FAO, Non-wood Forest Products Series No. 11, FAO, Rome, p. 158.

Hamilton A. 2003. *Medicinal Plants and Conservation: Issues and Approaches* online. UK, WWF [20th November 2003]. Portable Document Format.http://www.wwf.org.uk/filelibrary/pdf/medplantsandcons.pdf.

Ramakrishnappa K. 2002. Impact of Cultivation and Gathering of Medicinal Plants on Biodiversity: Case studies from India. In: *Biodiversity and the Ecosystem Approach in Agriculture, Forestry and Fisheries* online, FAO [cited 25 March 2003]. Available from Internet: http://www.fao.org/DOCREP/005/AA021E/AA021e00.htm

Schippmann U., Leaman D. J. and Cunningham A. B. 2002. Impact of Cultivation and Gathering of Medicinal Plants on Biodiversity: Global Trends and Issues. In: *Biodiversity and the Ecosystem Approach in Agriculture, Forestry and Fisheries.* FAO, 1-21.

Tabuti J. R. S., Lye K. A. and Dhillon S. S. 2003. Traditional herbal drugs of Bulamogi, Uganda: plants, use and administration. *Journal of Ethnopharmacology*, Vol. 88, No. 1: p. 19-44.

Trivedi P. C. 2004. *Medicinal Plants Conservation and Utilization,* Aavishkar Publishers, Jaipur.

Vandekop P., Alam G. and de Steenhuijsen Piters B. 2006. Developing a sustainable medicinal-plant chain in India. Ruben, R., Slingerland, M. and Nijhoff, H., eds. *Agro-food Chains and Networks for Development*, Springer, Netherlands.

Wakdikar S. 2004. Global health care challenge: Indian experiences and new prescriptions. *Electronic Journal of Biotechnology*, Vol. 7, No. 3: 214-220.

5

Molecular Marker Based Genetic Diversity Analysis of Two Economically Important Plants (*Zanthoxylum hamiltonianum* and *Nepenthes khasiana*) of North-East India

K. Medhi, B.P. Purohit, A. J. Gogoi, S.P. Saikia, P.B. Kanjilal, and B. S. Bhau

1. Introduction

Genetic variability refers to any variation in the nucleotides, genes, chromosomes, or whole genomes of organisms (the genome is the entire complement of DNA (deoxyribonucleic acid) within the cells or organelles of the organism). Genetic variation at its most elementary level is represented by differences in the sequences of nucleotides (adenine, cytosine, guanine and thymine) that form the DNA within the cells of the organism. Nucleotide variation is measured for discrete section of the chromosome and controls a particular characteristic of an organism.

Two spots were identified as 'Megadiversity Hot Spots' in India, the Northeastern Himalayas and the Western Ghats. But India as a whole has been marked as one of the 12 megadiversity regions of the world ofhaving the largest biovariability. As a part of Northeastern Himalaya's megadiversity hot spots, Northeast India is unique with profusion of habitats which features diverse biota with a high level of endemism. Physiographically, the region comprises of Eastern Himalayas, North East Hills (Patkai-Naga Hills and Lushai Hills), the Brahmaputra and Barak Valley. Forest area in Northeast India covers 64% of the total geographical area, but the area is decreasing at an alarming rate. During 1991 and 1999, about 1800 sq km of forest cover have been decreased. It is estimated worldwide that about 24 hectares of rain forest disappear every minute. India has 45,000 plant species with 15,000-20,000 proven to be of medicinal value. The National Medicinal Plant Board has identified 32 commercially important medicinal plants in Northeast India. Khasi-Jaintia Hills, Patkai-Lushai Hills, Cachar-Mikir Hills, Arunachal Pradesh and Sikkim are the areas among 24 areas identified in India as micro hotspots. The rich diversified flora of Northeast India is comprised of many valuable or economically important plants, 60% of which are used for medicinal values and some plants are for other uses.

Farmers are using the genetic variability in wild and cultivated plants for thousands of years to develop their crops and raise new breeds of livestock. Genetic variability gives species the ability to adapt to changing environments, including new pests, diseases and new climatic conditions. The total genetic variation or the genetic resource is the basis for more resilient production method against drought or overgrazing, hence it can reduce the potential of soil erosion and enables to cope with stresses such as world's rising human population, environmental problems (desertification, deforestation, erosion, etc.) and climate change particularly global warming that could bring about drastic changes in the location of the world's agro-ecological zones.

DNA-based molecular markers acted as versatile tools and found their position in various fields like taxonomy, physiology, embryology, genetic engineering etc. Ever since their development, they are constantly being modified to enhance their utility and to bring about atomization in the process of genome analysis.

A large number of methods exist for the assessment of genetic variability in plants. A combination of morphological traits and proteins profiling such as isozymes, allozymes and seed storage have conventionally been applied in a number of plant species. However such traits are influenced by different biotic and abiotic factors. Hence the result elucidated based on such studies do not provide a true measure of the genetic variability. DNA-based markers, being wholly genetic make up, are free from environmental modulations. RAPD markers have proved to be very useful tools in providing a convenient and rapid assessment of the genetic difference between genotypes (Williams *et al.* 1990).

2. Molecular Markers

Hybridization Based Markers

The classical DNA fingerprinting strategy relies on the hybridization of restricted digested genomic DNA with labeled multi-allelic, multi-locus probes. The technique of classical DNA fingerprinting is methodologically derived from Restriction Fragment Length Polymorphism (RFLP) analysis and is mainly distinguished from the later technique by the kind of hybridization probe applied to reveal polymorphism. To obtain a typical DNA fingerprint, probes are used to create complex banding pattern by recognizing multiple DNA loci simultaneously. Each of these loci is characterized by more or less regular array of tandemly repeated DNA motifs that occurs in different numbers at different loci.

Two categories of such multi-locus probes are mainly used:

1) Cloned DNA fragments or oligonucleiotides which are compliment to the so called 'minisatellites' i.e. tandem repeats of a basic motif of about 10-60 bp.

2) Exemplified by oligonucliotide probes which are complimentary to the so called 'simple sequence' or 'microsatellites' i.e. tandem repeats of very short motifs of about 1-5 bp.

RFLP is a reproducible and reliable co-dominant marker and is capable of detecting multiple alleles. It has been used extensively in construction of genetic maps, gene pyramiding through introgression and marker assisted breeding and selection (Mohan *et al.* 1997; Dean and Schmidt, 1995)

Microsatellite derived RFLP markers have been used by many workers like Serret *et al.* (1997). Its utilization is limited as it is expensive, labour intensive and technically complex as it involves radioactive probes for detection and a large amount of DNA for characterization of putative probes to be used in RFLP for genome analysis. Thus, this technique becomes impractical in analysis of genetic resources.

The hybridization based DNA fingerprinting includes a variety of steps:

- Complete digestion of genomic DNA with an appropriate Restriction Enzyme.
- Electrophoretic separation of the restricted fragments usually performed on agarose gels.
- Denaturation and blotting of the separated DNA fragments onto a membrane.
- Hybridization of the membrane to radioactively / non-radioactively labeled multi-locus probes.
- Detection of hybridizing fragments (i.e. fingerprints) by autoradiography or nonradioactive approaches.

PCR-Based Markers

With the invention of the polymerase chain reaction (PCR) by Mullis *et al.* (1986), DNA fingerprinting took huge strides forward in both discriminating power and ability to recover information from very small starting samples. PCR involves the amplification of specific regions of DNA using a cycling of temperature and a thermostable polymerase enzyme along with sequence specific primers of DNA.

Shortly after its introduction, many variants of the basic PCR strategy were developed and it was obvious that the technique would also be feasible for the detection of polymorphism. Initial attempts to reveal DNA polymorphism are made use of specific primers complementary to known sequences. These experiments are demonstrated that primers, which are complementary to flanking regions of minisatellites and microsatellites/ simple sequence loci, yield highly polymorphic amplification products.

Another strategy made use of semi-specific primers, which are complementary to repetitive DNA elements. For human genome analysis, an abundant class of randomly interspersed DNA called 'Alu repeats' was preferentially used for this purpose and Alu PCR revealed a considerable level of polymorphism. Later on the principle of Alu-PCR was adopted for other species as well.

Anonymous markers are defined by PCR amplification of anonymous DNA sequences from the genomic DNA using arbitrary or semi-arbitrary primers to generate complex fingerprint patterns (Weising, 1995).

Markers Based on Arbitrary Primers

During 1990s, several laboratories introduced a strategy that made use of one or two short GC rich primers of arbitrary sequence to generate PCR amplification products from genomic DNA. Based on arbitrarily primed PCR many techniques were evolved such as Randomly Amplified Polymorphic DNAs (RAPDs), Arbitrarily Primed PCR (AP-PCR) and DNA Amplification fingerprinting (DAF).

Randomly Amplified Polymorphic DNAs (RAPDs)

The RAPD procedure was first described by Williams *et al.* in 1990. It is technically the simplest variation of the arbitrarily primed PCR methods. Primers with 10 nucleotides and a GC content of at least 50% are generally used and are universal; this can be used for both prokaryotes and eukaryotes.

The RAPD has got many advantages like (1) no prior sequence knowledge is necessary; (2) small amounts of DNA can be used (25 nanograms per reaction); (3) the technique is technically simple; (4) detection is not radioactive; and (5) a large number of potential markers can be generated using readily available markers. The technique also reveals relatively high levels of polymorphism and has a low cost compared to other techniques. One of the problems associated with this technique is low annealing temperature that can cause some unspecific binding with primers binding to sequences that are not completely complementary.

Arbitrarily Primed PCR (AP-PCR)

AP-PCR was introduced by Welsh and McClelland in 1990. Oligonucleiotides of 20 or more nucleotides, originally designed for other purposes, were used as primers. Two cycles with low stringency were followed by 30-40 cycles with high stringency. Alpha (á) [^{32}P] dCTP was introduced in the last 20-30 cycles. Radio labeled products were separated by polyacrylamide gel electrophoresis and made visible by autoradiography. The AP-PCR variant of the arbitrary PCR method is used the least, compared to RAPD and DAF. It is also the most complicated and uses radioisotopes.

DNA Amplification Fingerprinting (DAF)

DAF was introduced by Caetano-Anolles *et al.* (1991a & b). The authors used short primers, often 5-8 nucleotides long, with either low or high stringency annealing steps and a two temperature instead of the standard three temperature cycle program. Resulting fragments were separated on polyacrylamide gels and visualized by silver staining.

Several limitations are there with RAPD, DAF and AP-PCR, despite of simplicity and ease to perform these techniques (Quiros *et al.* 1995; Jones *et al.* 1997). The reproducibility of these techniques are very poor with a slight changes in the reaction conditions like amount of template DNA, $MgCl_2$ concentration, thermocycler used and source of *Taq* polymerase.

Markers Based on Semi-Arbitrary Primers

AFLP (Amplified Fragment Length Polymorphism) and Microsatellite Primed PCR (MP-PCR) are the markers that use semi-specific primers based on either restriction sites (as in AFLP) or tentative microsatellite regions (as in MP-PCR).

Amplified Fragment Length Polymorphism

The AFLP technique introduced by Vos *et al.* (1995) is based on the selective PCR amplification of restriction fragments from a total digest of genomic DNA. The technique involves three steps:

(i) Restriction of the DNA and ligation of oligonucleotide adapters,

(ii) Selective amplification of sets of restriction fragments, and

(iii) Gel analysis of the amplified fragments.

PCR amplification of restriction fragments is achieved by using the adapter and restriction site sequence as target sites for primer annealing. The selective amplification is achieved by the use of primers that extend into the restriction fragments, amplifying only those fragments in which the primer extensions match the nucleotides flanking the restriction sites. Using this method, sets of restriction fragments may be visualized by PCR without knowledge of nucleotide sequence. The method allows the specific co-amplification of high numbers of restriction fragments. The number of fragments that can be analyzed simultaneously, however, is dependent on the resolution of the detection system.

The major advantage of the AFLP technique is the large number of polymorphisms that the method generates. Most applications of AFLP have been for gene mapping. Its ability to differentiate individuals in a population makes the technique useful for paternity analyses, gene flow studies and cultivar research. Other advantageous features of the AFLP technique are that no sequence information is required, the PCR technique is fast, results are reproducible, and standard kits are available (Wolfe & Liston, 1998). The major problem associated with this technique is that the number of steps needed to produce results involves two steps of PCR amplification which add an additional cost to the technique (Wolfe & Liston, 1998). Moreover the choice of primer may influence the number of bands amplified and the level of polymorphisms found.

Microsatellite Primed PCR (MP-PCR)

MP-PCR includes Simple Sequence Repeats (SSRs), short tandem repeats (STRs) or simple sequence length polymorphisms (SSLPs) where microsatellites are tandem repeats of sequence units generally less than 5 bp in length. They are thought to be produced by errors in DNA replication, when the DNA polymerase "slips" when copying in the repeats region, changing the number of repeats in the region.

Microsatellites appear to be hypervariable, in addition to which their co-dominance and reproducibility make them ideal for genome mapping, as well as for population genetic studies and diversity studies. Inter-SSRs (ISSRs), Direct amplification of minisatellite DNA markers (DAMD-PCR are the variant of the RAPD technique, as are chloroplast microsatellites (cpSSRs), which are similar to nuclear microsatellites but the repeat is usually only 1 bp.

The microsatellite protocol is simple, once primers have been designed. The first stage is a PCR, depending upon the method of detection one of the primers is fluorescently or radioactively labelled. The PCR products are separated on high resolution polyacrylamide gels, and the products detected with a fluorescence detector (for example an automated sequencer) or an X-ray film.

Sequencing-Based Markers

The hyper-variable microsatellite regions as mentioned earlier, may be converted into Sequence Tagged Microsatellite Sites (STMS). STS and STMS are PCR based co-dominant marker systems wherein the homozygote can be differentiated from heterozygote. Furthermore, in case of STMS, variable number of tandem repeats (VNTRs) present in different individuals can be detected as different sized amplification products that are essentially co-dominant

DNA sequencing can also be used as a definitive means for identifying species. Variations due to transversion, insertion or deletion can be assessed directly and information on a defined locus can be obtained. Genetic variation occurs extensively at the single nucleotide level. Direct sequencing can efficiently identify such single nucleotide polymorphisms that usually depend on how closely related are the organisms being compared. Some of the sequencing-based strategies are analysis of the variable internal transcribed spacer (ITS) sequences of ribosomal DNA (rDNA). The ITS region of 18s - 26s rDNA has proved to be a useful sequence for phylogenetic studies in many angiosperm families. The level of ITS sequence variation suitable for phylogenetic analysis is found at various taxonomic levels within families, depending on the linkage.

Sequence targeted markers are simple assays with higher allelic information owing to co-dominance. Therefore, they find their application basically in cultivar identification, genotyping and population studies.

Cleaved Amplified Polymorphic Sites (CAPS)

This is a PCR-RFLP based co-dominant marker system which involves amplification of genomic DNA on utilizing PCR primer synthesized on the basis of sequence information available in databank of genomic or cDNA sequences or cloned RAPD bands. CAPS is identified as an attractive molecular marker chiefly because of the flexibility which allows modulating the fragment size of the cleaved and uncleaved amplification products by adjusting the placement of CAPS primer. CAPS markers have been used in the selection of accessions resistant to tomato spotted wilt virus (TSWV) by Moury *et al.* (2000).

3. Application of Molecular Markers

Molecular markers are indispensable tools to understand the genetic structures of populations. In present day condition, reliable results of phylogenetics and taxonomic relatedness study can be obtained by using various molecular markers as these markers are not affected by the physiological and environmental conditions. Molecular markers are useful in identification and authentication of medicinal plants for the pharmaceutical industries.

A wide choice of genetic markers is now available to the conservation biologists, but each of these molecular markers exhibits different properties.

Table 1
Comparison table of a few molecular marker techniques

Markers	Level of Polymorphism	Locus specificity	Co-dominance of alleles	Reproducibility	Technical demands	Developmental cost	Quantity of DNA needed
RAPD	Medium	No	No	Low	Low	Low	Low
AFLP	High	No	No/yes	High	High	Low	Low
Microsatellite	High	Yes	Yes	High	Low-Medium	High	Low
ISSR	Medium	No	No	Medium-high	Low-Medium	Low	Low
Minisatellite	High	No/yes	No/yes	High	High	Medium-High	High

A similar kind of marker has been applied for two plant species, each having different economical value to assess their available variability within the population.

Case Study 1. Genetic diversity studies in *Zanthoxylum hamiltonianum* (*Tejmui*).

Scientific		Classification
Kingdom	:	Plantae
Division	:	Magnaliophyta
Class	:	Magnaliopsida
Sub class	:	Rosidae
Order	:	Sapindales
Family	:	Rutaceae
Genus	:	*Zanthoxylum*
Species	:	*hamiltonianum*

Zanthoxylum comes from the Greek words "xanthos" and "xylum" meaning, "yellow wood". It is a genus of about 250 species of deciduous and evergreen trees and shrubs in the family Rutaceae, native to warm temperate and subtropical areas worldwide. There are a few Species of *Zanthoxylum* available in Northeast India with high ethno medicinal uses. These are *Z. hamiltonianum, Z. rhetsa, Z. alatum, Z. oxyphyllum, Z. ovalifolium* etc.

Zanthoxylum is having medicinal properties, such as treating stomach ache, tooth ache, intestinal worms, rheumatism, scabies, snakebites, fever and cholera. The bark of several species of *Zanthoxylum* is medicinally active and noted for febrifugal, sudorific and diuretic properties. Several Indian species of *Zanthoxylum* have been used by the villagers and tribal since ages for treatment of a wide range of ailments. *Z. rhetsa* has anti-diarrhea activity, it is also shown antimicrobial and anti-inflammatory activities and it is used in intestinal cestode infection; *Z. oxyphyllum* has stimulant, stringent and digestive properties and used in dyspepsia and diarrhea; roots of *Z. hamiltonianum* serves as an insecticide capable of destroying mosquito larvae; these are to mention a few of the medicinal uses of highly medicinal and aromatic plants.

Apart from the numerous medicinal uses, *Zanthoxylum* is known to produce several chemical compounds such as flavonoids, terpenes and quinines. Several alkaloids have been isolated from the stem bark and root bark. The fruits of a number of its species yield essential oils. The oils from *Z. acanthopodium* and *Z. armatum* are potential sources of Linalool, an important perfumery on a limited scale for the production of war tar oil. *Z. hamiltonianum* (Tejmui) (Fig 1a), a large evergreen scandent shrub, armed with recurred prickles which on very old stems have more or less cylindrical and corky bases, young parts pubescent or velvety, pith small and not separate. Bark dry grey outside, cream coloured inside. Fruits are rich in oil (Fig. 1b)

Figures 1. a) *Z. hamiltonianum* (Tejmui) plant growing in nature, b) fruits of tejmui

About 38 samples were collected from the districts of Jorhat, Tinsukia, Golaghat, Karbi Anglong, and Sonitpur. Genomic DNA from tender leaves was isolated using modified CTAB method (Fig. 2) RAPD detected a very high level of polymorphism for the population of *Zanthoxylum hamiltonianum*. The Jaccard's similarity coefficient value ranged from 0.541 to 0.975. 9 primers out of 150 primers screened for polymorphism were selected.

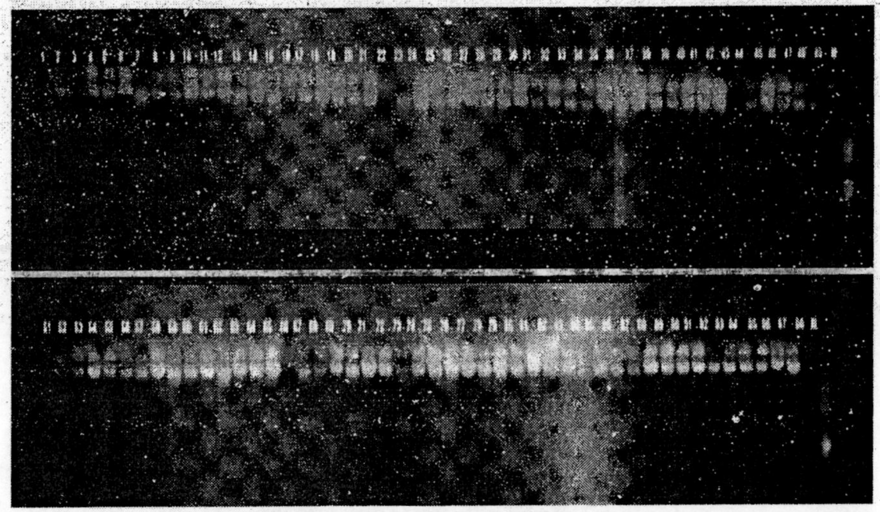

Figure 2. Isolated genomic DNA from Different *Zanthoxylum hamiltonianum* genotypes

Table 2
Optimization of PCR conditions for RAPD analysis technique in *Z. hamiltonianum*.

PCR Parameter	Tested Range	Optimum condition
DNA concentration (mg)	0.025, 0.05, 0.075	0.05
Primer (pmol)	5,10,15,20	10
dNTPs (mM)	0.2, 0.4, 0.6	0.4
Hot start Taq DNA polymerase)	0.13m, 0.17m, 0.25m	0.25m
$Mgcl_{2\,(mM)}$	1, 1.2, 1.6, 2.4	2.4
1% BSA (ml)	0.9,1.8,2.7	0.9
Annealing temperature	34°c, 35°c, 37°c	34^0 c
No. of PCR cycles	30, 35, 40	35

These 9 primers produce a total of 53 amplicons, out of which 32 were polymorphic. Primers like OPAA-18, OPBH-08 and OPC-18 were very much informative as each of them produces more than 70% polymorphism.

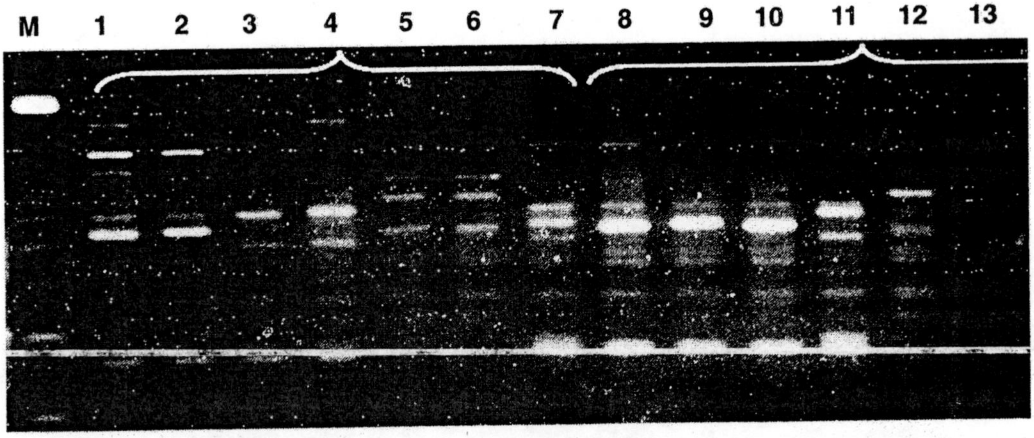

Figure 3. Primer testing in *Z. hamiltonianum*

The UPGMA cluster analysis permitted the discrimination of all the genotypes and their sorting into five broad groups. Thus from the RAPD data analysis the genetic divergence of the *Zanthoxylum hamiltonianum* was confirmed at DNA level. In fact, the greatest challenge of conservation is to preserve genetic variability in order to ensure the evolutionary processes (Soule *et al.* 1986), since in fragmented environment there is a tendency to lose inter-population gene exchange. As the seeds of the genus *Zanthoxylum* shows a very low degree of variability, thus, these studies might be helpful in preparing and planning future conservation strategies for this species.

Case study 2. Genetic diversity studies in *Nepenthes khasiana* (*Tiew-rakot, Ksete-phare, Memang-koksi*)

Scientific		Classification
Kingdom	:	Plantae
Division	:	Magnoliophyta
Class	:	Caryophyllales
Family	:	Nepenthaceae
Genus	:	*Nepenthes*
Species	:	*khasiana*

Nepenthaceae, a monotypic family with a single genus *Nepenthes* comprises roughly 117 species with numerous natural and many cultivated hybrids. The name *Nepenthes* comes from a Greek word "Nepenthe", a drug of forgetfulness mentioned in Greek mythology, meaning "the one that chases away sorrows" (Mukerjee *et al.* 1984).

The genus *Nepenthes* consist of more or less woody, sprawling or veining plants. They grow in loose, organic debris of the forest floor or sometimes in the crooks of tree branches and other similar places and thrive best in moist atmosphere where temperature ranges from 21°C to 30°C or a few degrees less in winter. They are evergreen dioeciously shrubs. *Nepenthes* have a passive trap and the pitcher is designed as such to ensure that the insect is attracted to a landing platform, the lid; and then passed on to the ribbed rim and to the inner surface, where it tumbles into the pool of digestive liquid and gets digested thus compensating for nitrogen and energy deficiency in the soil (Kitching & Schofield, 1986).

Figures 4a. A closed pitcher of *N. khasiana;* 4b. Opened lid pitcher

N. khasiana is a short, prostrate, undershrub with sub-cylindric pitcher is found in Garo, Khasi and Jaintia hills in Meghalaya upto 1,200 msl. The pitcher of the plant containing insects is rubbed into a paste and mixed with water and is given to cholera patients (Fig. 4a and 4b).

The species is officially classified as a threatened species and is included in the list of rare and threatened taxon of India (Jain & Baishaya, 1977; Jain & Sastry, 1980). It is also included in the list of Convention of International Trade on Endangered Species (CITES) of the world. (Mao & Kharbuli, 2002). It can be said that the species represent ancient endemic remnants of older floras which usually occur in land masses of geologically antiquity (palaeoendemic) (Bramwell, 1972) and hence, the species is of great botanical and horticultural interest. Due to consistent habitat destruction, the number of plants in Meghalaya forests has declined significantly (Fig. 5). The plant has also been reported for its ethno-medicinal uses. The fluid of unopened pitcher of *N. khasiana* is used by local Khasis and Garos as eye drops to cure cataract and night blindness (Kumar *et al.*1980). The unopened pitcher with its contents is made into a paste and applied for various skin diseases, including leprosy (Rao, 1981). The unopened pitcher fluid mixed with rice beer is taken to ease urinary troubles and blockages (Kharkongor & Joseph, 1981). The unopened pitcher fluid is also taken for treatment of stomach trouble, diabetes and female disease (Joseph & Joseph, 1986).

Figure 5. Habitat destruction of *Nepenthes khasiana* growing on the side of busy road in Tura (Garo Hills)

Forty eight (48) samples from six natural populations located in different parts of Meghalaya were selected for the evaluation of genetic diversity in *N. khasiana*. The collection places are Baghmara (BM), Jarain (JR), Rongsai (RSN), Baghmara Massighat (BMG), Chokpot (CP) and Bandari (BD). Genomic DNA extraction was carried out using

CTAB method (Doyle & Doyle, 1987) with some modifications viz. PVP was added to overcome the problem of phenolics in the leaf tissue of *N. khasiana.* Addition of PVP leads to the recovery of intact DNA. This method also yielded a colourless, high quality and quantity of genomic DNA. The DNA concentration was calculated using electrophorectic method. The genomics DNA recovered was not very intact. It has got a smear of small DNA fragment having different sizes. The concentration of genomic DNA isolated from lyophilized leaf tissue (100 mg each) ranged from 0.01 to 0.082 μg/μl (fig. 6). This DNA was dissolved in sterilized TE buffer and stored at 5°C. With all the samples, good quality and quantity of genomic DNA was obtained.

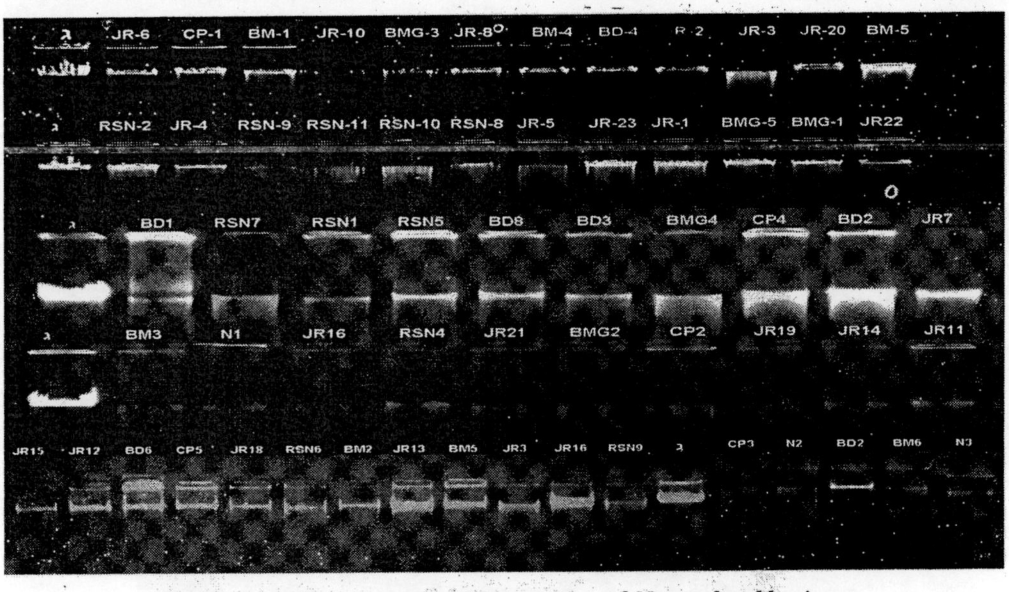

Figure 6. Isolated genomic DNA samples of *Nepenthes khasiana*

A total of 149 repeatable RAPD bands were generated under the optimized conditions (95°C for 3 mins; 95°C for1 min; 34°C for 2 mins; 72°C for 2 mins) for 35 cycles followed by 72°C for 10 min and kept on hold at 4°C which were obtained after altering various parameters during the standardization process. The Fig.7 represents RAPD profiles obtained from some of the primers like OPP-15.

A total of one hundred and thirty two bands (88.59%) were polymorphic and seventeen bands (11.41%) were monomorphic. The average number of bands was 12.4 per primer. Maximum polymorphism (18 bands i.e. 90%) was observed with primer OPP-15. Maximum numbers of monomorphic bands (4 bands) were observed with 0PC-11 out of 10 bands (40%). The average percentage of polymorphic bands for all samples per primers was 88.7%.

In this study, genetic discrimination of *N. khasiana* genotypes collected from different parts of Meghalaya was accomplished using the PCR based versatile RAPD technique. The RAPD results are concordant with other similar studies. These experiments demonstrated the reliability of RAPD analysis to detect DNA polymorphism and the relationship among the highly endangered and endemic *N. khasiana* plants. The wide variation in genetic distance among the different *N. khasiana* species revealed by RAPD techniques reflected a high level of polymorphism at the DNA level. Varying numbers of bands were amplified in different individuals with twelve primers. Zhou *et al.* (1998) studied the correlation between the number of RAPD loci investigated and the reliability of the information on genetic variation and found that stable information was obtained when the locus number exceeds 70. In the present study, a total of 149 amplification markers were generated, 132 of them were polymorphic.

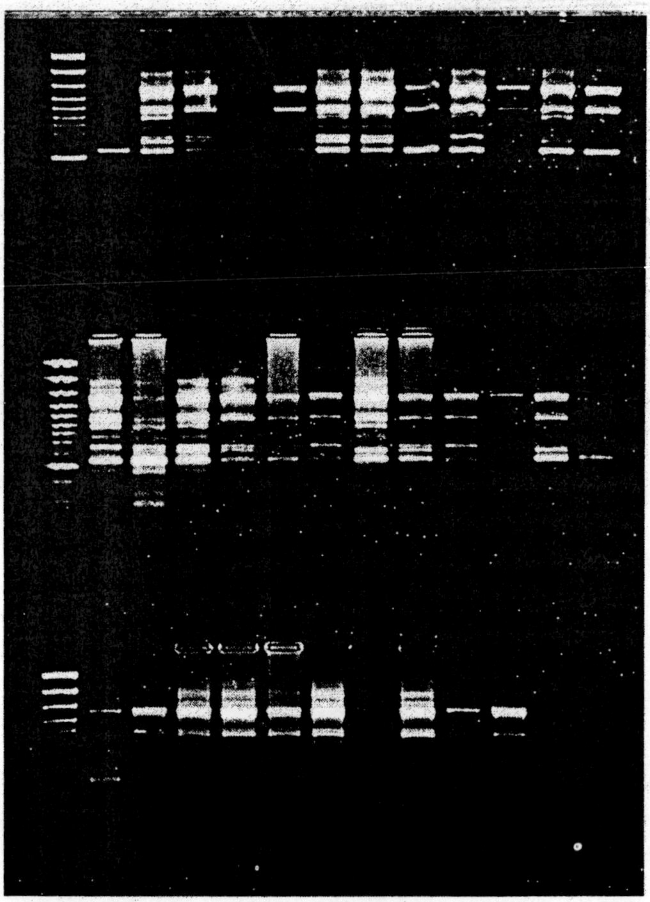

Figure 7. RAPD profiles obtained from some of the primers like OPP-15

4. Conclusion

Phylogeographic and population genetic studies of plant species, especially those that have experienced population declines and range reductions, could greatly benefit from this type of study. Increasing knowledge about the relationship between population size and plant fitness parameters is essential for conservation measures regarding rare, endangered and endemic plant species. Although this association is obvious in the case of *Zanthoxylum hamiltonianum* and *Nepenthes khasiana*, but further studies are needed to examine the underlying mechanisms and to separate clearly between genetic factors, environmental influences and probable pollinator shortage – including their interactions. However, the negative effects in both the plants found under study are small population size as have been demonstrated, conservation measures should be aimed at keeping natural populations sufficiently large enough to retain genetic diversity and minimize their risk of extinction. Molecular marker technology is now integrated into existing plant breeding programmes all over the world to allow researchers to access, transfer and combine genes at a rate with a precision not previously possible.

From the present RAPD study in *Zanthoxylum hamiltonianum*, it can be concluded that the samples collected from different parts of Assam have got considerable genetic variation. This is although the primary study of genetic diversity on the species of *Z. hamiltonianum*, and the number of markers scored is too low that can not give clear picture on exact genetic diversity, still it is suggested that there is good evolutionary growth among the population and it can be helpful in maintaining germ banks, with a view to maintain variability and contributing to the *in situ* conservation programs.

RAPD analysis study in *N. khasiana* can be concluded that the 48 samples collected from Meghalaya gave indication that they have got considerable genetic diversity. This genetic diversity can be utilized for further conservation programmes of *N. khasiana* plants. The prominent cultivars with highest degree of the diversity are the samples 24 and 48 (0.37). Some of other diverse samples of *N. khasiana* in our study are samples 4 and 48 (0.403), 1 and 42 (0.417), 31 and 48 (0.427) and 1 and 4 (0.429). As this is the first report on *N. khasiana* genetic diversity, this record needs further support by more studies of similar types.

This is the first study of genetic diversity on the germplasm of *N. khasiana* samples from different parts of Meghalaya and this work is going to be the diverging point for future conservation programs related to *N. khasiana* in maintaining its natural population and further protecting the habitats against any further loss on its population. Since it is the only species of *Nepenthes* found in India hence it is endemic and thus required to be conserved and maintained the biological diversity. Genetic diversity among the plants increases the chances of their survival rate against changing environmental factors.

The results of our study have shown polymorphism among the various populations, thus indicating a positive result in the conservation point of view. The greatest challenge of conservation is to preserve genetic variability in order to ensure the evolutionary processes, since in the fragmented environments there is a tendency to lose intrapopulation genetic variability due to low frequency of interpopulation gene exchange. These plants that are under a continuous threat due to various factors are regarded as endangered and endemic thus can be conserved by protecting their natural habitat and maintaining their natural population.

Acknowledgements: [We thank Dr. P.G. Rao, Director, NEIST, Jorhat, Assam for providing facilities and encouragement during the study. Thanks are also due to the Chief Conservator of Forests, Government of Meghalaya for his kind permission for *N. khasiana* exploration at various localities of the Khasi, Jaintia and Garo Hills. BSB, KM & PBK. We are also grateful to Department of Biotechnology for funding of *Zanthoxylum* project].

REFERENCES

Caetano-Anolles G., Bassam B. J. and Gresshoff P. M. 1991a. DNA amplification fingerprinting using very short arbitrary oligonucleotide primers. *Biotechnology* 9: 553-557.

Caetano-Anolles G., Bassam B. J. and Gresshoff P. M. 1991b. DNA amplification fingerprinting: A strategy for genome analysis. *Plant Molecular Biology Reporter* 9 (4): 294-307.

Dean C. and Schmidt R. 1995. Plant genomes: A current molecular description. *Annu Rev Plant Mol Biol* 46: 395-418.

Doyle J. J. and Doyle J. L. 1987. A rapid DNA isolation procedure for small quantities of fresh leaf tissue. *Photochemical Bulletin*. 19: 11-15.

Jain S. K. and Baishya A. K. 1977. *Nepenthes khasiana*: An endangered species. *Hornbill*: 17-18.

Jain S. K. and Sastry A.R.K. 1980. *Threatened Plants of India—A State of the Art Report*. Botanical Survey of India, Calcutta, 48p.

Jones C. J., Edwards K. J., Castaglione S., Winfield M. O., Sala F., van de Wiel C., Bredemeijer G., Vosman B., Matthes M., Daly A., Brettscneider R., Bettini P., Buiatti M., Maestri E., Malcevschii A, Marmiroli N., Aert R., Volckaert G., Rueda Linacero R. J., Vazquez A. and Karp A. 1997. Reproducibility testing of RAPD, AFLP and SSR markers in plants by a network of European laboratories. *Mol Breed* 3: 381-390.

Joseph J. and Joseph K. 1986. *Insectivorous Plants of Khasi and Jaintia Hills, Meghalaya, India*. Botanical Survey of India. Calcutta.

Kharkongor P. and Joseph J. 1981. Folklore medico-botany of rural khasi and Jantia tribes in Meghalaya. In Jain SK (Ed.). *Glimpses of India Ethnobotany,* Oxford & IBH, New Delhi, 124-136.

Kitching R. L. and Schofield C. 1986. Every pitcher tells a story. *New Scientist* 109: 48-50.

Mao A. and Kharbuli P. 2002. Distribution and status of Nepenthes khasiana hook. F. – A rare endemic plant of Meghalaya. *Phytotaxonomy* 2: 77-83.

Mohan M., Nair S., Bhagwat A., Krishna T. G., Yano M., Bhatia C. R. and Sasaki T. 1997. Genome mapping, molecular markers and marker assissted selection in crop plants. *Mol Breed* 3: 87-103.

Moury B., Pflieger S., Blattes A., Lefebvre A. and Palloix A. 2000. A CAPS marker to assist selection of tomato spotted wilt virus (TSWV) resistance in pepper. *Genome* 43 (1): 137-142.

Mukherjee A., Dam D.P. and Daen P. 1984. Pitcher plant (An ornamental climber of Meghalaya). Botanical Survey of India. *India Horticulture* (April-June): 16.

Mullis K., Faloona S., Scharf S., Saiki R., Horn G. and Erlich H. 1986. Specific enzymatic amplification of DNA in vitro: the polymerase chain reaction. *Cold Spring Harbor Symp Quant Biol* 51: 263-273.

Quiros C. F., This P., Laudie M., Benet A., Chevre A. M. and Delseny M. 1991. Analysis of a set of RAPD markers by hybridization and sequencing in *Brassica*: A note of caution. *Plant Cell Reports* 14: 630-634.

Serret M. D., Udupa S. M. and Weigand F. 1997. Assessment of genetic diversity of cultivated chickpea using microsatellite derived RFLP markers: Implications for origin. *Plant Breeding* 116: 573-578.

Vos P., Hogers R., Bleeker M., Reijans M., van de Lee T., Hornes M., Freijters A., Pot J., Peleman J., Kuiper M. and Zabeau M. 1995. AFLP: a new technique for DNA fingerprinting. *Nucl Acids Res* 23: 4407-4414.

Weising K., Nybom H., Wolff K. and Meyer W. 1995. *DNA Fingerprinting in Plants and Fungi.* CRC Press, Boca Raton, Fla.

Welsh J. and McClelland M. 1990. Fingerprinting genomes using PCR with arbitrary primers. *Nucl Acids Res* 18: 7213-7218.

Williams J.G.K., Kubelik A.R., Levak K.J., Rafalski J.A. and Tingey S.V. 1990. DNA polymorphism amplification by arbitrary primers are useful as genetic markers. *Nucleic Acids Res.,* 18: 6531-6535.

Wolff K., Zeitkeiwicz E. and Hofstra H. 1995. Ientification of Chrysanthemum cultivars and stability of DNA fingerprint patterns, *Theor Appl Genet* 91: 439-447.

6

Intellectual Property Rights and Traditional Knowledge

D. Marngar

1. Introduction

Intellectual Property Rights (IPRs)

"IPR is defined by the World Intellectual Property Organization (WIPO) as the exclusive rights given to people over the creations of their minds". They usually give the creator an exclusive right over the use of his/her creations for a certain period of time. Or "The rights resulting from intellectual activities" like Literary, artistic, scientific works, performances or performing artists, phonograms and broadcasts; inventions in all fields of human endeavor; Scientific discoveries; Industrial designs; Trade marks; Service marks and commercial names and designations; All other rights resulting from intellectual activity in the industrial, scientific, literary and artistic fields.

Objectives of IPRs

- The law of intellectual property allows people to own their creativity and innovation in the same way that they can own physical property.
- It allows you to claim ownership of the story, play or song you have written down, the picture you have painted, the invention you have drawn or made, the logo or product you have designed, the music, poem or dance that you have performed and recorded.
- Intellectual property is the physical manifestation of your ideas.

Understanding traditional knowledge (TK)

Traditional knowledge (TK) is a broad term. We refer to TK as the original rights of indigenous people and local communities over various elements — plants and genetic resources, traditional medicines, agricultural methods, local technologies, and cultural products (e.g. weaving, pottery, poetry, folklore, music and the like) which they have discovered and developed.

Traditional knowledge can be viewed as a kind of collective sense within shared communities. For example, the application and efficacy of medicinal plants as understood by traditional medical practitioners in a tribal community forms a specialized island of knowledge and wisdom. This knowledge underlying a society's art, medicine and agriculture may be handed down from generation to generation via give-and-take collaboration and the practice of apprenticeship.

However, since many indigenous communities do not have a written tradition or culture, recognition of TK through a system of IPRs has always been difficult and complex.

Objectives of the paper

- An overview of TK
- Impact of IPRs on TK
- Challenges confronting TK holders
- Legal protections of TK
- Key issues & recommendations on the role of IPR on TK

2. Overview of TK and IPR

Originality of TK

Traditional knowledge (TK), indigenous knowledge (IK), and local knowledge generally refer to the matured long-standing traditions and practices of certain regional, indigenous or local communities. Traditional knowledge also encompasses the wisdom, knowledge and teachings of these communities. In many cases, traditional knowledge has been orally passed for generations from person to person. Some forms of traditional knowledge are expressed through stories, legends, folklore, rituals, songs and even laws. Other forms of traditional knowledge are often expressed through different means.

Indigenous and local communities justly cherish traditional knowledge (TK) as a part of their very cultural identities. Maintaining the distinct knowledge systems that give rise to TK can be vital for their future well-being, sustainable development and for their intellectual and cultural vitality. For many communities, TK forms part of a holistic world-view, and is inseparable from their very ways of life and their cultural values, spiritual beliefs and customary legal systems. This means that it is vital to sustain not merely the knowledge but the social and physical environment of which it forms an integral part.

Challenges for Strengthening of TK

TK holders face various difficulties relating to the very survival of the knowledge, as the cultural survival of communities is under threat. External social and environmental pressures, migration, the encroachment of modern lifestyles and the disruption of traditional ways of life can all weaken the traditional means of maintaining or passing knowledge on to future generations. There may be a risk of losing the very language that gives the primary voice to a knowledge tradition and the spiritual world-view that sustains this tradition. Thus, a primary need is to preserve the knowledge that is held by elders and communities throughout the world:-

- To find ways of strengthening and nurturing the roots of TK, even in times of social dislocation and change, so that the fruits of TK can be enjoyed by future generations and so that traditional communities can continue to thrive and develop in ways consistent with their own values and interests.
- TK holders stress that their TK should not be used by others inappropriately, without their consent and arrangements for fair sharing of the benefits.
- To call for greater respect and recognition for the TK values, contributions and concerns of TK holders.

Examples

- A recent agreement would give traditional healers in Samoa a share of the benefits from a new AIDS drug drawing on their knowledge of the *mamala* tree.
- The Kani tribe of South India is to share the benefits from a new sports drug that is based on their knowledge of the medicinal plant *Trichopus zeylanicus*.
- Traditional ecological knowledge held by Aboriginal communities in Canada has proven to be valuable in environmental planning and resource management
- For some communities, TK provides a pathway to social and economic development and more culturally appropriate forms of tourism: the Seri people of Mexico use the *Arte Seri* mark to distinguish their craftworks based on their TK and associated genetic resources and to support a sustainable trade in these products.
- In 2001, China granted more than 3000 patents on innovative developments within the field of Traditional Chinese Medicine.

Issues established to promote TK

The promotion of TK is important for communities, particularly in developing and under developed countries. TK plays an important role in the economic and social life of these countries. TK helps strengthening cultural identity and enhanced use of such knowledge to achieve social and developmental goals, such as sustainable agriculture, public health and conservation of biodiversity. Developing and under developed countries are implementing international agreements that associated with TK and the use of genetic resources is promoted and disseminated, thus their national interests are safeguarded.

- World Intellectual Property Organization (WIPO) started to work on TK in 1998 - 1999 and conducted fact-finding missions in 28 countries by meeting directly to TK holders, learning of the needs and expectations of some 3,000 representatives of TK-holding communities, academicians, researchers and private sector representatives in sixty locations around the world. Their insights and perspectives still guide WIPO's work.
- The WIPO Intergovernmental Committee (IGC) on Intellectual Property and Genetic Resources, Traditional Knowledge and Folklore was established in 2001 as an international policy forum to capacity building and pooling of practical experiences in this complex area.
- In 1978, the World Health Organization (WHO) first recognized the relevance of traditional medicine as a source of primary health care in the Primary Health Care Declaration of Alma Ata. The topic has been addressed since 1976 by the WHO Traditional Medicine Team, including through the development of the WHO Traditional Medicine Strategy.
- The Declaration on the Rights of Indigenous Peoples in 1993 has been under negotiation. The draft refers to the entitlement of indigenous peoples in relation to their cultural and intellectual property.

Misutilization of TK

Many important technologies being derived directly from both the codified (documented) as well as non-codified (not documented but may be orally transmitted) TK, but others seek benefit from TK (especially for industrial or commercial advantage), but the knowledge has been misappropriated and the role and contribution of TK holders has not been recognized and respected. The indigenous knowledge has always been an easily accessible treasure and thus has been susceptible to misappropriation. The indigenous traditional knowledge, particularly, related to the treatment of various diseases has provided the leads for development of biologically active molecules by the technology rich countries. In other words, traditional knowledge is being exploited for bio-prospecting. TK is misappropriated, also because it is conveniently assumed that since it has been publicly disclosed; communities have given up all claims over it.

Protection of TK

TK systems are so closely intertwined with the cultural identity of many diverse communities. Effective protections may be found in a coordinated "menu" of different options for protection. This could perhaps be underpinned by an internationally agreed set of common objectives and core principles that could form part of the international legal framework. The key is to provide TK holders with an appropriate choice of forms of protection, to empower them to assess their interests and choose their own directions for the protection and use of their TK, and to ensure there is adequate capacity to carry through protection strategies. TK protections include:

- One and the only form of protection, is the application of laws to prevent unauthorized or inappropriate use of TK by third parties beyond the traditional circle.
- Protection on the use of knowledge such as traditional technical know-how, innovations, information practices, skills, learning of TK systems such as traditional agricultural, environmental or medicinal knowledge, traditional cultural expressions (TCEs) or expressions of folklore (such as songs, chants, narratives, motifs and designs).
- To respect the holistic context of TK and TCEs, and to recognize the linkages between these aspects of life and culture of traditional communities (example, the same body of customary law is likely to apply to both TK and TCEs).

Outcomes of TK protection

1. Protection of TK should principally benefit the holders of the knowledge particularly
   - The indigenous and traditional communities.
   - Peoples that develop, maintain and identify culturally with TK and seek to pass it on between generations,
   - To recognize individuals within these communities and people.
2. TK often provides researchers with a lead to isolate valuable active compounds within biological resources. Such genetic and biological resources are linked to TK and traditional practices through the utilization and conservation of the resource, which has often occurred over generations and through their common use in modern scientific research.

3. The protection of TK is often closely linked to protection of biodiversity, in particular under the Convention on Biological Diversity (CBD).

3. Impacts of IPR on TK

A. Negative Impacts

Intellectual Property Rights Devalue Indigenous Knowledge

In agriculture and in health care, humans have been the biodiversity experts. In partnership with plants, third world people meet the needs for food and medicine in their societies. Farmers are the seed keepers and seed custodians in most peasant societies- they select, store, sow and conserve seed. Healers are the knowledge providers of plant-based medicine. According to an ethnobotanical survey, there are more than 7500 species used as medicinal plants by the indigenous medical traditions of India. These traditions are kept alive by 360 740 Ayurvedic practitioners, 29 701 Unani experts and 11 644 specialists of Siddha, in addition to millions of housewives, birth attendants and herbal healers using village based health traditions. Seventy percent of health care needs in India are still based on traditional systems of medicines. Eighty percent of seeds used by farmers still come from farmers' seed supply. India is thus still a predominantly biodiversity based economy and farmer's knowledge is very central to this economy.

Since time immemorial relationship between people and plants is very congenial, but this is now being ruptured through patent regimes being expanded to cover plants, animals and microorganisms.

Intellectual property rights (IPRs) are an instrument in the GATT agreement which will dispossess rural farmers of their power, control and knowledge. IPRs in GATT and other international platforms aim to take seed out of peasant's custody and medical plants out of the hands of local healers and make it the private property of TNCs (transnational corporations). By adding 'trade related' to IPRs, GATT has forced issues of the ownership of genetic resources and life forms on to the agenda of international trade through Trade Related Intellectual Property Rights (TRIPs).

In fact, most bodies of traditional knowledge will produce minimal economic benefit from applying IPRs. Instead, the promotion of IPRs is often an excuse to bring outlying markets into the fold of an unfamiliar economic model by which the developed economies exert control through shear size and precedent.

Furthermore, because of their continuing unequal statures, developing countries bear the lion's share of misappropriation of traditional knowledge. But the true loss is not the pennies which their indigenous populations might earn selling herbs and local seeds. The true price is induction into a system of trade where the deck is already stacked in favor of existing and highly developed proprietary interests.

Instead of treating traditional knowledge as a marketable commodity or property in its own right, greater value may be forged using collective knowledge as the capital to fund the building blocks of higher-order economic processes.

IPRs: The Privatization of Biodiversity and Biodiversity Related Knowledge

The thrust of the western IPR regimes in the area of biodiversity is diametrically opposed to indigenous knowledge systems. Its knowledge is considered to be the product of individual creativity. In this paradigm IPRs represent the property rights to the products of mind, thereby resulting in knowledge and creativity being so narrowly defined that the creativity of nature and non-western knowledge systems have been ignored.

The two categories of IPRs that have a direct impact on the erosion of prior rights of communities are patents and plant breeders' rights. Plant breeders' rights negate the contribution of Third World farmers as breeders and hence undermine farmers' rights. Patents allow the usurpation of indigenous knowledge as a western invention through minor tinkering or trivial translation.

A frequent comment heard in scientific and lay circles, is that "we should patent all our traditional knowledge and biodiversity." However, neither traditional knowledge nor biodiversity can be patented by indigenous practitioners because for indigenous societies, it is not 'novel', it is ancient. The reason that the collective and cumulative innovation of millions people in thousands of years can be 'pirated' and claimed as an 'innovation' of western trained scientists or corporations, is because of two reasons. The first reason is the colonial hangover of the idea that science is unique to the west and indigenous knowledge systems cannot be treated as scientific.

The second reason is that countries like the US, where most pirated indigenous innovations are filed for patenting, do not recognise the existing knowledge of other countries as prior art. Thus, while patent regimes offer no protection to indigenous communities for their common innovation and their common resources, they allow the appropriation of their biodiversity and knowledge by scientists and commercial interests of other cultures, including members of the 'modern' scientific culture in their own societies.

IPRs: An Instrument of Piracy

IPR regimes in the context of 'free trade' and 'trade liberalisation' become instruments of piracy at three levels:

Resource piracy in which the biological and natural resources of communities and the country are freely taken, without recognition or permission and are used to build up global economies. For example, the transfer of basmati varieties of rice from India to build up the rice economy of the US; the free flow of neem seeds from the farms, fields and commons to corporations like W. R. Grace for export.

Intellectual and cultural piracy in which the cultural and intellectual heritage of communities and the country is freely taken without recognition or permission and is used for claiming IPRs such as patents and trademarks even though the primary innovation and creativity has not taken place through corporate investment. For instance, the use by US corporations of the trade name 'basmati' for their aromatic rice, or Pepsi's use of the trade name 'Bikaneri Bhujia'.

Economic piracy in which the domestic and international markets are usurped through the use of trade names and IPRs, thereby destroying local economies and national economies where the original innovation took place and hence wiping out the livelihoods and economic survival of millions. For example US rice traders usurping European markets; Grace usurping the US market from small scale Indian producers of neem based biopesticides.

B. Positive Impacts

Giving legally recognized ownership of knowledge to communities through IPRs has several benefits. It will raise the profile of that knowledge and encourage respect for it both inside and outside the knowledge holding communities. This will enthuse the younger members of such communities to contribute to the further development of that knowledge. Furthermore, prospects of economic returns for the use of that knowledge by others will act as a further incentive for the community members to respect their knowledge and continue to engage in practices in that knowledge. Prior protection will also provide disclosure, use and proliferation of such knowledge, which might have otherwise been eroded.

The local communities or individuals do not have the knowledge or the means to safeguard their property in a system, which has its origin in very different cultural values and attitudes. The communities have a storehouse of knowledge about their flora and fauna – their habits, their habitats, their seasonal behaviour and the like – and it is only logical and in consonance with natural justice that they are given a greater say as a matter of right in all matters regarding the study, extraction and commercialization of the biodiversity. A policy that does not obstruct the advancement of knowledge and provides for valid, sustainable use and adequate intellectual property protection with just benefit sharing is what we need.

To date, debate on IPRs and biodiversity has focused on patents and plant breeders' rights. Provisions under undisclosed information or trade secrets could be invoked to protect traditional knowledge not available in the public domain. Geographical indications and trademarks could also be the alternative tools for indigenous and local communities seeking to gain economic benefits from their traditional knowledge. The potential value of geographical indications and trademarks is in protecting plants and germplasms that are specific and unique to geographical regions. They could protect and reward traditions while allowing innovation. They will emphasize the relationships between human cultures and their local land and environment. They are not freely transferable from one owner to another. They can be maintained as long as the collective tradition is maintained.

New experiments are beginning to emerge on benefit sharing models for indigenous innovation. An experience in India is worth sharing. It relates to a medicine that is based on the active ingredient of a plant, *Trichopus zeylanicus*, found in the tropical forests of southwestern India and collected by the Kani tribal people. Scientists at the Tropical Botanic Garden and Research Institute (TBGRI) in Kerala learned of the plant, which is claimed to bolster the immune system and provide additional energy, while on an expedition with the Kani in 1987. These scientists isolated and tested the ingredient and incorporated it into a compound, which they christened "Jeevani", the giver of life. The tonic is now being manufactured by a major Ayurvedic drug company in Kerala. In 1995, an agreement was struck for to share the license fee and 2% of sales of the product as royalty, that was receivable by TBGRI, will be shared on a fifty-fifty basis with the tribe. This marks perhaps the first time that for IP held by a tribe, compensation in the form of cash benefits has gone directly to the source of the IP holders. We are still on the learning curve in this experiment but we need to multiply such examples globally.

4. Challenges Confronting TK Holders

TK holders face various difficulties. In some cases, the very survival of the knowledge is at stake, as the cultural survival of communities is under threat. External social and environmental pressures, migration, the encroachment of modern lifestyles and the disruption of traditional ways of life can all weaken the traditional means of maintaining or passing knowledge on to future generations.

There may be a risk of losing the very language that gives the primary voice to a knowledge tradition and the spiritual world-view that sustains this tradition. Either through acculturation or diffusion, many traditional practices, associated beliefs and knowledge has been irretrievably lost. Thus, a primary need is to preserve the knowledge that is held by elders and communities throughout the world.

Another difficulty facing TK holders is the lack of respect and appreciation for such knowledge. For example, when a traditional healer provides a mixture of herbs to cure a sickness, the healer may not isolate and describe certain chemical compounds and describe their effect on the body in the terms of modern biochemistry, but the healer has, in effect, based this medical treatment upon generations of clinical trials undertaken by healers in the past, and on a solid empirical understanding of the interaction between the mixture and human physiology. Thus, sometimes the true understanding of the value of TK may be overlooked if its scientific and technical qualities are considered from a narrow cultural perspective. In fact, many consumers in western countries are turning to treatments based on TK, on the understanding that such "alternative" or "complementary" systems are soundly based on empirical observation over many generations.

Yet another problem confronting TK holders is the commercial exploitation of their knowledge by others, which raises questions of legal protection of TK against misuse, the role of prior informed consent and the need for equitable benefit-sharing. Cases involving natural products all bear evidence to the value of TK in the modern economy. A lack of experience with existing formal systems, limited economic resources, cultural factors, lack of a unified voice, and in many cases, a lack of clear national policy concerning the utilization and protection of TK. At the same time, the lack of understanding and clear rules concerning the appropriate use of TK creates areas of uncertainty for those seeking to use TK in research and development of new products.

There is a common need for well-established, culturally appropriate and predictable rules both for the holders and legitimate users of TK. A further challenge is to address the international dimension of the protection of TK and benefit-sharing for associated genetic resources, while learning from existing national experiences. Only through the participation of communities and countries from all regions can this work goes forward to produce effective and equitable outcomes that are acceptable to all stakeholders. These challenges are diverse and far reaching, and involve many areas of law and policy, reaching well beyond even the most expansive view of intellectual property.

In general, the preservation and protection against loss and degradation of TK should work hand-in-hand with the protection of TK against misuse and misappropriation. So when TK is recorded or documented with a view to preserving it for future generations, care needs to be taken to ensure that this act of preservation doesn't inadvertently facilitate the misappropriation or illegitimate use of the knowledge.

5. Legal Protection and Other Forms of Protection TK

A comprehensive strategy for protecting TK should consider through community, national, regional and international dimensions. The stronger the integration and coordination between each level, the more effectiveness is the TK protection. Many communities, countries and regional organizations are working to address these levels respectively. National laws are currently the prime mechanism for achieving protection and practical benefits for TK holders. For instance, Brazil, Costa Rica, India, Peru, Panama, the Philippines, Portugal, Thailand and the United States of America have all adopted *sui generis* laws that protect at least some aspect of TK (*sui generis* measures are specialized measures aimed exclusively at addressing the characteristics of specific subject matter, such as TK).

Protection should reflect the aspirations and expectations of TK holders and should promote respect for indigenous and customary practices, protocols and laws as far as possible. Several *sui generic* measures, as well as conventional IP law, have recognized elements of such customary law within a broader framework of protection. Economic aspects of development need to be addressed and the effective participation by TK holders is also important, in line with the principle of prior informed consent. TK protection should also be affordable, understandable and accessible to TK holders. The view is widely voiced that holders of TK should be entitled to fair and equitable sharing of benefits arising from the use of their knowledge. The international legal framework, within and beyond the IP system, is also an important consideration. Where TK is associated with genetic resources, the distribution of benefits should be consistent with measures established in accordance with the Convention on Biological Diversity (CBD), providing for sharing of benefits arising form the utilization of the genetic resources. Other important international instruments include the International Treaty on Plant Genetic Resources for Food and Agriculture of the Food and Agriculture Organization (FAO), the International Union for the Protection of New Varieties of Plants (UPOV), and the UN Convention to Combat Desertification (UNCCD). Other areas of international law, notably human rights and cultural policy, are also part of the context for protection of TK.

Two key demands on the IP system in particular have arisen in policy debate:

1. The call for recognition of the rights of TK holders relating to their TK, and,
2. Concerns about the unauthorized acquisition by third parties of IP rights over TK. Two forms of IP-related protection have therefore been developed and applied:

Positive protection: giving TK holders the right to take action or seek remedies against certain forms of misuse of TK; and

Defensive protection: safeguarding against illegitimate IP rights taken out by others over TK subject matter.

Positive Protection – Recognition of IP Rights in TK

Stakeholders have stressed that these two approaches should be undertaken in a complementary way. A comprehensive approach to protection in the interests of TK holders is unlikely to rely totally on one form or the other.

The kind of objectives that TK protection is intended to serve include:

- Recognition of value and promotion of respect for traditional knowledge systems
- Responsiveness to the actual needs of holders of TK
- Repression of misappropriation of TK and other unfair and inequitable uses
- Protection of tradition-based creativity and innovation
- Support of TK systems and empowerment of TK holders
- Promotion of equitable benefit-sharing from use of TK
- Promotion of the use of TK for a bottom-up approach to development

The International Policy Framework

In 1992, the United Nations Conference on Environment and Development was held in Rio de Janeiro under the auspices of UNEP and led to the Rio Declaration on Environment and Development establishing the Convention on Biological Diversity (CBD) to promote the conservation of biological diversity, the sustainable use of its components and the fair and equitable sharing of benefits arising out of the utilization of genetic resources. Provisions on the respect and recognition of TK are a key element of the CBD.

1994, the UN Convention to Combat Desertification (UNCCD) provided for the protection of traditional knowledge in the ecological environment as well as the sharing of benefits arising from any commercial utilization of this TK

In 1989 the FAO Conference recognized Farmers' Rights. In 2000, FAO Conference adopted the International Treaty on Plant Genetic Resources for Food and Agriculture which provides recognition of farmers' rights, including "the protection of traditional knowledge relevant to plant genetic resources for food and agriculture"

In 2000, the United Nations Conference on Development (UNCTAD) in its Plan of Action stressed the importance of studying ways to protect traditional knowledge, innovations and practices of local and indigenous communities and enhance cooperation on research and development on technologies associated with the sustainable use of biological resources.

The United Nations Development Programme (UNDP) conducts extensive capacity building work on TK, including on aspects of legal protection and equitable benefit-sharing.

In 2000, the WIPO Intergovernmental Committee on Intellectual Property and Genetic Resources, Traditional Knowledge and Folklore addressing both policy and practical linkages between the IP system and the concerns of practitioners and custodians of traditional knowledge. The Committee is moving towards an international understanding of the shared objectives and principles that should guide the protection of TK.

The Doha Declaration adopted, at the Doha WTO Ministerial Conference, in 2001, instructed the TRIPS Council to examine *inter alia* the protection of TK and folklore

Work on traditional knowledge protection is also continuing elsewhere in the United Nations system. For instance, the United Nations University has issued a report on "The Role of Registers and Databases in the Protection of TK".

Existing IPR Laws

The policy debate about TK and the IP system has underlined the limitations of existing IP laws in meeting all the needs and expectations of TK holders. Even so, existing IP laws have been successfully used to protect against some forms of misuse and misappropriation of TK, including through the laws of patents, trademarks, geographical indications, industrial designs and trade secrets. Certain adaptations or modifications to IP law may be needed to make it work better. For example, TK is often held collectively by communities, rather than by individual owners – this is often cited as a drawback in protecting TK. Yet it is possible to form associations, community corporations or similar legal bodies to act on behalf of the community. In some countries, government agencies take active role acting in trust for the community. Some forms of protection, such as remedies against unfair competition and breach of confidence, do not require specific right holders. There are also concerns that the cost of using the IP system is a particular obstacle for TK holders. This has led to explore capacity building, evolution of legal concepts to take greater account of TK perspectives, the use of alternative dispute resolution and a more active role for government agencies and other players. Existing IP rights have been used in the following ways:

Unfair competition and trade practices laws: these allow for action to be taken against false or misleading claims that a product is authentically indigenous, or has been produced or endorsed by, or otherwise associated with, a particular traditional community. For instance, a company has been legally barred from describing various hand made products as "certified authentic" and "Aboriginal art" when they were not painted by Aboriginal people and had not undergone any certification process.

Patents: when practitioners innovate within the traditional framework, they have been able to use the patent system to protect their innovations. For example, in 2001 China granted 3300 patents for innovations within the field of Traditional Chinese Medicine. Systems have been developed to ensure that illegitimate patent rights are not granted over TK subject matter that is not a true invention.

Distinctive signs (trade marks, collective marks, certification marks, and geographical indications): traditional signs, symbols and terms associated with TK have been protected as marks, and have been safeguarded against third parties' claims of trade mark rights. For instance, the Seri people of Mexico, faced with competition from mass production, registered the *Arte Seri* trademark to protect authentic ironwood products that are produced by traditional methods from the *Olneya tesota* tree. Conservation of this unique species of tree was also a fact or in protecting the trademark.

The law of confidentiality and trade secrets: this has been used to protect non-disclosed TK, including secret and sacred TK. Customary laws of communities often require that certain knowledge be disclosed only to certain recipients. Courts have awarded remedies for breach of confidence when such customary laws are violated.

Nationals' Legal Protection of TK

Organization of African Unity: African Model Legislation for the Protection of the Rights of Local Communities, Farmers and Breeders, and for the Regulation of Access to Biological Resources of 2000

Peru was established by Law No. 27, 811 of 2002, whose objectives are to protect TK, to promote fair and equitable distribution of benefits, to ensure that the use of the knowledge takes place with the prior informed consent of the indigenous peoples and to prevent misappropriation.

Philippines: Indigenous Peoples Rights Act of 1997.

Portugal: - Portugal's *sui generic* Decree-Law No. 118, of April 20, 2002 is the registration, conservation and legal custody of genetic resources and TK. The Law provides protection against the "commercial or industrial reproduction and/or use" of TK developed by local communities, collectively or individually.

Thailand: Act on Protection and Promotion of Traditional Thai Medicinal Intelligence, B.E 2542.

Costa Rica: The Biodiversity Law No. 7788 of Costa Rica aims at regulating access to TK. It provides for the equitable distribution to TK holders of the benefits arising from the use of TK.

Andean Countries: Decision 486 of 2000 on a Common Intellectual Property Regime

China: Patent Law of 2000 Related database/register: Traditional Chinese Medicine Patent Database.

India: The Patent Amendments Act of 2005 Related database/register: Traditional Knowledge Digital Library.

Regulatory Frameworks for TK

EC:

European Directive 2004/24 as Regards Traditional Herbal Medicinal Products

European Directive 2006/509 on Traditional Specialities Guaranteed Related database/register: Register on Traditional Knowledge in Agricultural Products and Food - Austria

India: Drugs and Cosmetics Act, 1940

Italy: Attuazione della direttiva 92/73/CEE in materia di medicinali omeopatici

Philippines: Traditional and Alternative Medicine Act of 1997.

Other Legal Concepts for the Protection of TK

When policymakers explore suitable legal mechanisms to protect TK against misappropriation, they consider a broader range of legal concepts apart from the kind of exclusive rights used in most forms of IP law. Several of these alternative concepts are briefly described as:

Prior Informed Consent (PIC): According to the principle of prior informed consent (PIC), TK holders should be fully consulted before their knowledge is accessed or used by third parties and an agreement should be reached on appropriate terms; they should also be fully informed about the consequences of the intended use. The agreed scope of use may be set out in contracts, licenses or agreements, which would also specify how benefits arising from the use of the TK should be shared. The principle of PIC concerning access to genetic resources is one of the cornerstones of the CBD. Given the close relationship between genetic resources and some forms of TK, this same principle is also used in a number of national laws concerning access to and use of TK.

Equitable benefit-sharing: The idea of an equitable balancing of interests is common to many legal systems. In IP law, this is often phrased in terms of a balancing of the interests of right holders and the general public. The fair and equitable sharing of benefits from the use of genetic resources is one of the objectives of the CBD, and the CBD also encourages equitable sharing of benefits from the use of certain forms of TK. Thus, the principle of equitable benefit-sharing is found in a number of national laws governing access and use of TK, especially when TK is associated with genetic resources. According to this principle, the TK holders would receive an equitable share of the benefits that arise from the use of the TK, which may be expressed in terms of a compensatory payment or other non-monetary benefits. An entitlement to equitable benefit-sharing may be particularly appropriate in situations where exclusive property rights are considered inappropriate.

E.g.: Traditional healers of ***Samoa*** were recently acknowledged in a benefit-sharing agreement concerning the development of prostratin, an anti-AIDS compound derived from the Samoan native *mamala* tree (*Homalanthus nutans*). Prostratin forces the HIV out of reservoirs in the body, thus allowing anti-retroviral drugs to attack it. The bark of the *mamala* has been used by traditional healers to treat hepatitis, among other medicinal uses of the tree. This traditional knowledge guided researchers in their search for valuable therapeutic compounds. Reportedly, revenues from the development of prostratin will be shared with the village where the compound was found and with the families of the healers who helped discover it. Revenues will also be applied to further HIV/AIDS research. It is also proposed to license the prostratin research to drug makers so that the resultant drugs are made available to developing countries for free, at cost, or at a nominal profit.

Unfair competition: this is defined as "any act of competition contrary to honest practices in industrial or commercial matters", and includes various acts that mislead the public or cause confusion. Unfair competition law has been used as a potential basis for protection of integrated circuit layout designs, geographical indications, undisclosed information and test data, and phonograms. It has also been discussed and used as a potential legal basis for protecting TK against various forms of unfair commercial use.

Respect for customary laws and practices: Customary laws, protocols and practices often define how traditional communities develop, hold and transmit TK. For example, certain sacred or secret TK may only be permitted to be disclosed to certain initiated individuals within an Indigenous community. Customary laws and practices may define custodial rights and obligations over TK, including obligations to guard it against misuse or improper disclosure; they may determine how TK is to be used, how benefits should be shared, and how disputes are to be settled, as well as many other aspects of the preservation, use and exercise of knowledge. For example, in North America, the inheritance and transfer of "medicine bundles" within or between families is accompanied by the transmission of traditional medical knowledge and certain rights to practice, transmit and apply that knowledge. The ownership of the physical bundle is often attached to exclusive rights to exploit the products and processes associated with the TK that the bundle signifies.

TK holders have called for their customary laws, practices and beliefs to be recognized and respected by those seeking to use their TK. For many representatives of traditional communities, this is a cornerstone of appropriate forms of protection. This has led to consideration of a range of ways of respecting customary laws and practices within other legal mechanisms, including within conventional IP systems.

Defensive Protection and Safeguarding Against Illegitimate IP Rights Over TK

Defensive protection can be valuable and effective in blocking illegitimate IP rights, but it does not stop others from actively using or exploiting TK. Some form of positive protection is needed to prevent unauthorized use. This is why a comprehensive approach to protection needs to consider positive and defensive protection as two sides of the same coin. For instance, publishing TK as a defensive measure may block others from patenting that TK, but it can also make the knowledge more accessible and put it in the public domain – this can, ironically, make it easier for third parties to use the knowledge against the wishes of the TK holders. The main focus of defensive protection measures has been in the patent system.

Defensive protection aims at ensuring that existing TK is not patented by third parties – ideally, by ensuring that relevant TK is fully taken into account when a patent is examined for its novelty and inventiveness. Normally, a claimed invention in a patent application is assessed against the so called *"prior art"* – the defined body of knowledge that is considered relevant to the validity of a patent. For example, if TK has been published in a journal before the applicable date of a patent application. it is part of the relevant prior art, and the application cannot validly claim that TK as an invention – the invention would not be considered novel. In recent years, concern has been expressed that TK should be given greater attention as relevant prior art, so that patents are less likely to cover existing publicly disclosed TK. Defensive protection of TK has two aspects:

- ***A legal aspect***: how to ensure that the criteria defining relevant prior art apply to the TK – for example, this could mean ensuring that orally disclosed information must be taken into account (since many important bodies of TK are normally transmitted and disseminated by oral means)

- **A practical aspect**: how to ensure that the TK is actually available to search authorities and patent examiners, and is readily accessible – for example, TK and defensive protection against the turmeric patent

Two US based Indians Suman K Das and Hari Har P. Cohly were granted a US Patent 5,40,504 on March 28, 1995 on *"Use of Turmeric in wound healing"*. The patent was assigned to University of Mississippi Medical Center, USA. This patent claimed the administration of an effective amount of turmeric through local and oral route to enhance the wound healing process, as a novel finding. Any patent, before it is granted has to fulfil the basic requirements of novelty, non-obviousness and utility. Thus, if the claims have been covered by relevant published art, then the patent becomes invalid. CSIR could locate 32 references (some of them being more than one hundred years old and in Sanskrit, Urdu and Hindi), which showed that this finding was well known in India prior to filing of this patent. The formal request for re-examination of the patent was filed by CSIR at USPTO on 28th October 1996.

The first office action in the re-examination was issued by USPTO on 28th March, 1997, which rejected all the six claims based on the references submitted by CSIR as being 'anticipated by the submitted references' and therefore considered invalid under 35 U.S.C. 102 and 103. In the second Office action, the examiner rejected all the claims once again and made his action final on 20th November 1997 as anticipated and obvious. The re-examination certificate was issued on this case on April 21, 1998 bringing the reexamination proceedings to a close.

6. Bridges the Gap Between TK and Modern Scientific Knowledge

Traditional Knowledge Digital Library (TKDL)

The Traditional Knowledge Digital Library (TKDL): It is a database with a tool/ mechanism to understand the prior art available for Indian Systems of Medicine (Ayurveda / Siddha / Unani / Yoga) which is a codified knowledge. It is not a diagnostic/ usage database or TKDL is also not the prior art.

For Example, prior art of Indian Systems of Medicine in the form of text, comprises of the following main components:

(a) Plants, minerals or animals and their products thereof, as main ingredients.

(b) Method of preparation of a given formulation.

(c) Usage, e.g. in a particular disease or to maintain health.

TKDL, gives modern names to Plants / Diseases / Processes, mentioned in the literature related to Indian Systems of Medicine and establishes relationship between traditional knowledge and modern knowledge. It makes available the information existing in local languages into English, French, Spanish, German and Japanese in patent application format, easily understandable by patent examiners. Thus, TKDL database which breaks the language and format barrier is a tool which will provide defensive protection to rich traditional knowledge of India.

TKDL is an initiative of several Indian Government agencies, proposes to document the disclosed traditional medicinal knowledge available in public domain by sifting and collating information on TK from the existing disclosed literature covering Ayurveda. The TKDL compiles the information in digitized format in five international languages (English, German, French, Japanese and Spanish). An inter-disciplinary team of Ayurveda experts, a patent examiner, information technology experts, scientists and technical officers have worked for one and a half years for creating the TKDL of Ayurveda. TKDL seeks to give recognition and legitimacy to the existing TK and enable protection of such information from getting patented. Traditional Knowledge Resource Classification (TKRC) is an innovative structured classification system for the purpose of systematic arrangement, dissemination and retrieval. This has been developed for about 5000 subgroups against one group in international patent classification (IPC), i.e. AK61K35/78 related to medicinal plants. TKDL is a collaborative project between the Indian National Institute of Science Communication and Information Resources and the Department of Indian Systems of Medicine and Homeopathy, Ministry of Health and Family Welfare.

TKDL is linking to internationally accepted International Patent Classification (IPC) System which means building the bridge between the knowledge contained in an old Sanskrit Shloka and the computer screen of a patent examiner in Washington. This will eliminate the problem of granting wrong patents since the Indian rights to that knowledge will be known to the examiner.

It could act as a bridge between the traditional and modern knowledge systems. Availability of this knowledge in a retrievable form in many languages will give a major impetus to modern research in the developing world, as it itself can then get involved in innovative research on adding further value to this traditional knowledge; an example being the development of an allopathic medicine based on a traditional plant based therapeutic. Sustained efforts on the modernization of the traditional knowledge systems of the developing world will create higher awareness at national and international level and will establish a scientific approach that will ensure higher acceptability of these systems by practitioners of modern systems and public at large.

7. Issues and Recommendations

1. Regulate access to biological and genetic resources for our people's benefits.

2. Complement regulations with an uptodate inventory or registration system of TK.
3. Acknowledge the role of TK holders in improving TK and the rights of indigenous communities over TK.
4. Strengthen IPR systems to recognize the more informal, communal system of innovation.
5. Bridging the gap between the developed (scientific systems) and underdeveloped countries on common balance interests.

1. Laws regulating access to genetic and biological resources must be formulated. Restricting to access is the only way to gain benefits from our genetic and biological resources and to know more knowledge about these resources and use these knowledge to improve our ability to deliver industrial, agricultural or health services and thus improve the lives of our people. Today, we know that there has been a growing trend for natural foods, herbal medicines and natural supplements, in the world market. The Indian's rich bio-diversity offers various opportunities for local research activities in the area of plant breeding and in the development of biological derivatives and herbal preparations using our commonly known plant varieties. For example, herbal preparations using *Potentila* or *Lynniang* have been tested to have therapeutic effects. This should be where investments be placed as regards to biological and genetic resources.
2. Regulations with an uptodate inventory or registration system of TK. It would be difficult to impose regulations unless we have a national or state registry that identifies particular species. Thus, there should be a systematic documentation and inventory of plant and genetic resources and knowledge originating from indigenous and local communities shall become the basis for IPR protection.
3. Acknowledge the role of farmers and the rights of indigenous communities over TK. To ensure farmer's privilege to save, use, exchange, and sell products. For this purpose, policy makers will have to be better educated about their traditional practices, the enforceability of restrictions and the redistributive implications of the benefits vis-à-vis farmers and innovators. Under the context of globalize trading regime, it is important that such a system of intellectual property rights protection be established to safeguard the interest of our TK holders and protect the plant varieties against infringement and bio-piracy.
4. We should strengthen IPR systems to recognize the more informal, communal system of innovation. Intellectual property systems has always been criticized to cater only

on the dominant model of innovation and that no recognition is given for the protection of traditional knowledge, innovations and practices of indigenous communities although it is widely acknowledged that without the input of indigenous knowledge many products used extensively worldwide would not exist today. Our laws have failed to address this need. Perhaps, there is a need for a law that will provide for a separate *sui generis* system for the protection of community intellectual property rights.

5. To mitigate this problem, steps have to be taken to create a Traditional Knowledge Digital databases on traditional medicines and systems which will also lead to a Traditional Knowledge Resource Classification (TKRC). Linking this to internationally accepted International Patent Classification (IPC) System will mean building the bridge between the knowledge contained in an old TK and the modern scientific knowledge. This will eliminate the problem of hesitation and uncertainty within the scientific community. Eventually the creation of digitalized databases in the developing world would serve a bigger purpose in providing and enhancing its innovation capacity. It could integrate widely scattered and distributed references on the traditional knowledge systems of the developing world in a retrievable form. Thus it could act as a bridge between the traditional and modern knowledge systems.

8. Conclusion

The call for protection of TK against misuse or misappropriation raises deep policy questions and practical challenges alike. The changing social environment and the sense of historical dislocation, that currently affect many communities may actually strengthen and resolve to safeguard TK for the benefit of future generations. Just as the technological value of TK is increasingly recognized and its potential realized, the challenge is to ensure that the intellectual and cultural contribution of traditional communities is appropriately recognized. This means taking greater account of the needs and expectations of TK holding communities concerning the intellectual property system. Its traditional qualities and frequent close linkage with the natural environment mean that TK can form the basis of a sustainable and appropriate tool for locally-based development. It also provides a potential avenue for developing countries, particularly least-developed countries, to benefit from the knowledge economy. It requires respect for the values and concerns of traditional communities, as well as consideration of the full international policy and legal context, including a range of current international debates. Even new or expanded forms of IP protection would be inadequate to meet all the needs and expectations that have been voiced, but various forms of IP mechanism have been found to be practically useful. The current WIPO process aims at distilling the practical and policy lessons of a wide range of experience in many countries, with a view to building a shared policy perspective and effective practical tools. The WIPO work is framing the core principles that should underpin the protection of TK. This offers a potential foundation for international legal development in the form of precise policy and legislative options for enhanced protection of TK through adapted or expanded conventional IP systems or through stand-alone *sui generis* systems. This may in turn facilitate further development of an international consensus on the more detailed aspects of protection, as the lessons of practical experience in achieving these principles are better understood and shared. This should lead to strengthened linkages between the needs and interests of traditional communities and the core public policy principles of the IP system.

REFERENCES

Alexander Merle., Chamundeeswari K., Kambu Alphonse., Muller., Manuel Ruiz., Tobin. and Brendan. 2004. *The Role of Registers and Databases in the Protection of Traditional Knowledge*. Institute for Advanced Studies (UNU/IAS), United Nations University, Tokyo.

Blakeney Michael. 1997. Bioprospecting and the Protection of Traditional Medical Knowledge of Indigenous People: an Australian Perspective, in *E.I.P.R.*, 29(6): 298-303.

Blakeney Michael. 2000. The Protection of Traditional Knowledge under Intellectual Property Law, in *E.I.P.R.*, 22(6): 251-261.

Craig Donna G. 2005. Biological resources, intellectual property rights and international human rights: Impacts on indigenous and local communities. In: Burton Ong (ed): *Intellectual Property and Biological Resources*. National University of Singapore -Faculty of Law - Pacific Centre for Environmental Law / Singapore Academy of Law / Intellectual Property Office of Singapore / IP Academy. Marshall Cavendish Academic, Singapore.

Drahos Peter. 2000. Indigenous Knowledge, Intellectual Property and Biopiracy: Is a Global Biocollecting Society the Answer, in *E.I.P.R.* 22(6): 245-250.

Dutfield Graham. 2005. Legal and Economic Aspects of Traditional Knowledge, in *International Public Goods and Transfer of Technology Under a Globalized Intellectual Property Regime*, edited by J.H. Reichman & K. Maskus, Cambridge.

Dutfield Graham. 2001. TRIPS-related Aspects of Traditional Knowledge, in *33 Case W. Res. J. Int'l L.* 233.

Gibson Johanna. 2005. *Community Resources: Intellectual Property, International Trade, and Protection of Traditional Knowledge*. Ashgate Publishing Ltd., Aldershot.

Gopalakrishnan N.S. 2005. TRIPS and Protection of Traditional Knowledge of Genetic Resources: New Challenges to the Patent System, in *E.I.P.R.* 27(1): 11-18.

Gupta and Anil Kumar. 2004. *The Role of Intellectual Property Rights in the Sharing of Benefits Arising from the Use of Biological Resources and Traditional Knowledge*. World

Intellectual Property Organization (WIPO) / United Nations Environment Programme (UNEP). World Intellectual Property Organization (WIPO), Geneva,

Hardison and Preston D. 2005. *The Report on Traditional Knowledge Registers (TKRs) and Related Traditional Knowledge Databases (TKDBs).* Ad Hoc Open-Ended Inter-Sessional Working Group on Article 8(J) and Related Provisions of the Convention on Biological Diversity, Fourth meeting. UNEP/CBD/WG8J/4/INF/9. Secretariat of the Convention on Biological Diversity (CBD), Montréal.

Haugen and Hans Morten. 2005. Traditional knowledge and human rights. *The Journal of World Intellectual Property* 8(5): 663-678.

http://www.wipo.int/tk/en/tk/index.html.

Permanent Forum on Indigenous Issues, International Workshop on Free, Prior and Informed Consent and Indigenous Peoples, New York, 17-19 January 2005, PFII/2005/WS.2/5, United Nations Department of Economic and Social Affairs - Division for Social Policy and Development - Secretariat of the Permanent Forum on Indigenous Issues, New York, 2005.

"Protecting and Promoting TK: Systems, National Experiences and International Dimensions", UNCTAD, 2004

Secretariat of the Convention on Biological Diversity (CBD), *Convention on Biological Diversity Handbook,* 3rd ed. Secretariat of the Convention on Biological Diversity (CBD), Montréal, 2005.

Tauli-Corpuz, Victoria, *Biodiversity, Traditional Knowledge and Rights of Indigenous Peoples.* Tebtebba Foundation / Third World Network, Penang, Malaysia, 2004 (also available as: Permanent Forum on Indigenous Issues: Workshop on Traditional Knowledge, the United Nations and Indigenous Peoples, 21-23 September 2005, Panama City, Panama, PFII/2005/WS.TK/5. United Nations Department of Economic and Social Affairs - Division for Social Policy and Development - Secretariat of the Permanent Forum on Indigenous Issues, New York, 2005.

WHO Traditional Medicine Strategy 2002-2005 (WHO/EDM/TRM/2002.1) *The Role of Registers and Databases in the Protection of Traditional Knowledge – A Comparative Analysis,* United Nations University, Institute of Advanced Studies (UNU-IAS)

"WIPO-UNEP Study on the Role of Intellectual Property Rights in the Sharing of Benefits Arising from the Use of Biological Resources and Associated Traditional Knowledge", WIPO Publication No. 769E

"WIPO Technical Study on Patent Disclosure Requirements Related to Genetic Resources and Traditional Knowledge", WIPO Publication No. 786E "Consolidated Analysis of the Legal Protection of Traditional Knowledge"

WIPO Secretariat, "Intellectual Property Needs and Expectations of Traditional Knowledge Holders: WIPO Report on Fact-Finding Missions on Intellectual Property and Traditional Knowledge (1998-1999)" WIPO Secretariat, "Protection of Traditional Knowledge: Overview of Policy Objectives and Principles" (WIPO/GRTKF/IC/7/5)

7

A Precise Report of the Study on the Legume, *Dolichos biflorus* (L): An Ancient Herbal Medicine

Dibyendu Chakraborty and Jagatpati Tah

1. Introduction

Dolichos biflorus (Linn.) or *Macrotyloma uniflorum* [(Lamp) (Verdc)] commonly known as horsegram or kulthi is well known as the poor man's pulse since ancient times. It has gained popularity for base cropping on fallow and less fertile lands. It is highly drought resistant and short duration type, a valuable source of protein, calcium, iron and vitamin in animal feeds. It contains more than 22% seed protein which is greater than in gram and peas. It cures kidney trouble, eliminates stones (Calcium oxalate) and dissolves fat. It also helps to balance the billirubin level and sterol amount in the blood. In India it is grown on an area of 204 thousand hectares with a total production of 7074 thousand tonnes which gives an average yield of 350 kg/ha. The main reasons for its low yield are:

i) Susceptibility to high moisture content in soil and atmosphere; indeterminate growth; asymmetrical flowering and fruiting; high rate of flower fall; no dormancy; and pod shattering resulting to seed loss.

ii) Susceptibility to disease and insect pests; and

iii) Lack of adaptation to full package of practices involved.

Some of the morphological and functional characters of variability, germination, growth, flowering and maturity show asymmetrical trends. These require genetic investigations for identifying the responsible genes and their reaction to environmental fluctuations, major and minor elements of soil and some of the common stress factors, although it grows fairly well in saline and dry areas. The plant germinates under a wide range of temperature from 8 - 40°C with almost no dormancy; it requires a critical photoperiod and temperature when completed its vegetative growth by the minimum period of 45-50 days till the 2nd week of November, if it was grown in the last week of October. It has been observed that if it is grown earlier and just after the rainy season i.e. during the 1st week of September it shows prolonged vegetative phase and flowers after 75-80 days which is till the 1st week of December, when day length becomes shorter than the earlier months. However, during vegetative phase an optimum day length of 11hs followed by 9hs is required for flowering and seed maturity. Thus allowing the plants to grow in longer photoperiods during its vegetative phase in September often makes it to remain in vegetative phase during the rest of short day months and flower only in April when the days appear longer, hence, showing a life span not less than 8 months.

The growth of legume is sensitive to high moisture and soil water condition during summer months; thus, flowering and seed fertility are considerably affected and minimizing seed yield leading to shattering of pods, seed and seed germination. But in spite of these constraints and a longer flowering photoperiod, it cannot be suitably adapted in rainy summer months. Thus at laboratory condition, a varietal reaction to atmospheric moisture and soil water condition that prevails during the months of July-December was investigated.

Habitat: The legume is saline and acid resistant; erect to sub-erect. A good number of promising cultivars have been evolved in other states but introduction in West Bengal has failed due to high humidity and lack of basic information on genetic, breeding, botanical, and physiological characteristics, varietal response to humidity, soil, water, day length, rain, up and low lands and a viable cropping technology. However, often it is grown as a crop on fallow lands and rain fed dry upland locations and occasionally cultivated by the local people in the lateritic districts of West Bengal. Considering its adaptive characteristics, it is expected that the legume will be adapted to the rain fed drought prone upland locations.

Medicinal Use

Plant Part Use	Disease
Root	Hypertension
Stem	Blood sterol
Leaf	Jaundice
Seed	Kidney stone, Liver cirrhosis, and Regulation of menstrual cycle.

Genetic peculiarities: With uniform seed fertility in the dry months pronounced variations in the number of pods per unit area was recorded. Higher seed fertility was observed in some of the varieties grown under high humid conditions.

2. Review of Literature

Indeed, there is no report on GEI in *Dolichos biflorus* L and in fact this legume could not be included in the list of the agronomically important pulses in the crop calendar, because the legume is now grouped as a minor one and under-utilized and exploited one. Moreover, legume breeding work began in 1962 (Baluch & Soomro, 1968) and minor pulse like *D. biflorus* L failed to draw attention in research till recently.

With regards to this legume, so far only two studies on genetic parameters have been reported (Chatterjee, 1970; Singh & Meherdiratta, 1970; & Bhaumik, 1971).

In India, investigations were carried out on different intercropping and rotational systems at various locations during the years 1972-73 and 1973-74 (Mahapatra, *et al.* 1975). The chemical composition of the oil derived from the seeds of *Dolichos biflorus* L. was studied by Prakash & Purohit (1972). Molecular weight and sub-units of the lectin of *Dolichos biflorus* L. was studied by Pere *et al.* (1974). They showed that the end group had only alanin in the N-terminal position suggesting that the 4-subunits are identical. Aubery *et al.* (1973) observed the changes in chick embryonic cell growth that induced by a specific lectin, *Dolichos biflorus* L. Isolation characterization, and sub-unit structure of multiple forms of *Dolichos biflorus* L. lectin were observed (Carter & Etzler, 1975). Tah & Dasgupta (1988) studied its environmental interaction.

A study of the development and distribution of *Dolichos biflorus* L. was made by Talbot and Etzler (1978). The mutagenic effects of gamma rays and ethyl methane sulphonate in horse gram (*Dolichos biflorus* L.) have been reported by Kulkarni, *et al.* (1978). Preliminary studies on mutagenic effects of dimethyl sulphonate on *Dolichos biflorus* L. was reported by Kulkarni, *et al.* (1978). Studies on germination and seedling growth in M_1 of Hebbal-Hurali-1 indicated that the LD-50 for dimethyl sulphate treatment was about 0.03% (Kulkarni, *et al.* 1978).

The inheritance of seed colour and purple color on standard petal in horse gram (*Dolichos biflorus* L.) was analysed by Kulkarni, *et al* . (1978). The presence of black mottling on the testa; and of a purple eye on the standard are conditioned by single, independent and dominant genes designated by B^m and P_e, respectively.

In *Dolichos biflorus* L. the stability problem has not been investigated earlier. However, Sen, *et al.* (1965), Tomar, *et al.* (1973), Gupta & Ramanujan (1974a) studied the stability behavior in a similar legume chick pea. Singh & Mehra (1980) reported stability for the seed size character.

3. Materials and Methods

Location

The position of the location and physio-chemical characteristics of the Crop Research Farm where the research experiments was carried out is as follows:

Characteristics	CRF, Tarabag, Deptt of Botany, B.U.
1. Latitude	23° 14' 55" to 23° 15' 20" N
2. Longitude	83° 50' 35" to 87° 50' 54" E
3. Soil texture	Sandy clay
4. Soil pH	6.8
5. Conductivity (TSS) (mili-com-mhos/cm) Rating	0.12
6. Organic carbon (%) Rating	0.38
7. Available phosphorus(P)kg/ha	74.5 Medium
8. Available potassium(K)kg/ha	141.4 Medium Low

Material

During the 1st year, in 1994 genetic screening was undertaken on a collection of 51 accessions of the legume, *Dolichos biflorus* L. to ascertain breeding potential of the individual lines with regard to the yield attributes, yield of individual plant and their morphological behavior. The accessions were collected from (i) National Bureau of Plant Genetic Resource, Indian Council of Agricultural Research, Pusa Campus, New-Delhi-110 012, India; (ii) Regional Station of the National Bureau of Plant Genetic Resource at Vellanikera, Trichur-680 654, Kerala, India and (iii) P. K. V. Campus, Akola - 444 104, Maharastra, India. To homogenize the populations the accessions were grouped by seed coat colour and grown separately in successive years to develop pure lines. The colour-wise selections of the 51 pure breeding accessions were as follows:

Sl.No	Accession	Seed coat colour
1.	IC 23508	White
2.	PLKU 160	Black
3.	PLKU 514	Dark brown
4.	PLKU 74	Brownish white
5.	PLKU 180	Black pigmented

Sl.No	Accession	Seed coat colour
6.	PLKU 62B	Blackish white
7.	PLKU 128	Blackish
8.	PLKU 439	Light black
9.	IC 23495	Brownish
10.	IC 23509*	Brown
11.	PLKU 486	Blackish
12.	PLKU 150A*	Brown
13.	IC 4224*	Brown
14.	IC 1944	Dark brown
15.	IC 24411	Brownish
16.	IC 45225	Brownish white
17.	S 432*	Grayish brown
18.	IC 44011	Brownish
19.	S 436*	Whitish brown
20.	IC 23754	White
21.	IC 18531**	Brown
22.	PLKU 127	Blackish
23.	IC 15228**	Blackish brown
24.	IC 44012	Brownish
25.	IC 44013	Brownish
26.	IC 24842*	Brown
27.	PLKU 150IA	Brown
28.	114/82-17	Black
29.	IC 23507	Brown
30.	94/84-1	Brownish black
31.	S 528*	Brown
32.	S 448*	Black
33.	IC 44022	Brown
34.	IC 23467-2*	Grayish white
35.	IC 44020	Brown
36.	IC 23497*	Brown
37.	IC 23469*	Gray

(Contd...)

Sl.No	Accession	Seed coat colour
38.	IC 33309	Brownish
39.	PLKU 501**	Whitish brown
40.	IC 32746	Brownish
41.	IC 40720	Brownish
42.	IC 45702	Brownish
43.	5/83-19	Brownish
44.	S 439*	Blackish
45.	S 455**	Brown
46.	IC 44021	Brown
47.	IC 1527*	Brown
48.	S 526*	Brown
49.	IC 15227**	Brown
50.	PLKU 533*	Blackish
51.	IC 19430	Brown.

* Accessions which selected for details study after 1st adaptability trial.

** Accessions which were selected for diallele breeding programme.

Layout & Design

The land on which the experiment was carried out was divided into 3 blocks of same size and shape and each of the blocks into 51 plots of same size and shape to accommodate the 51 accessions in the experiment. Each plot consisted of three rows, 4.2 cm long with a row spacing of 30 cm accommodating 20 plants in each plot; the plot dimensions being 60 cm X 4.2 cm giving plant to plant and row to row a distance of 20 x 20 and 30 x 30 cm respectively. The two extreme rows of a plot and the two extreme plants of each row were discarded, thus giving one experimental row out of three rows per plot. Usual agronomical practices were followed to ensure proper growth of the crop. The accessions were subjected to grow under 6 environmental conditions represented by the months (i) June, (ii) July, (iii) August, (iv) September (v) October and (vi) November. Identical layout design was provided for each environment.

Characters, Studies and Recording of Data

20 characters were studied during experimentation which have been exhibited in the following tables:

Model

Following Perkins and Jinks (1968a, 1968b, 1973), Pooni & Jinks 1985 model and notations the additive effects were represented as the di values γiΣδΣ(u – v)d for each line and ej values as [1 + Σ(u – v)β] ej for each environment. Each γij forming the line I, environmental interaction component as γij or γij – (μ + di) as the (di – gij) or the γij values were obtained as the dependent one. These values were employed for the regression analysis. Hence, the regression model followed a linear expansion or γij = μ + di + εj + gij + δij where, m represented grand mean over all lines and environments; di as additive effects of the ith genotypes; ej as additive environmental contribution of the jth environment; gij as the genotype environmental contribution interaction of the ith line in jth environment, as shown separately in the analysis of variance. This was considered as a system following linear model of fixed effect of Σ_i di – Σ_j εj = εδij – Σ_j gij = 0.

Following the specifications above, each observation is linearly expanded. For example, observation 11 or γ_{11} is expanded to $\mu + d_1 + \varepsilon_1 + g_{11}$ where, $\mu = \gamma.. / t_s$; $\varepsilon_1 = \mu$ - Υ/s; $\varepsilon_j = \mu$ - Υ.1/t; gij is γij - μ - ε_1 which is generalized as gij = γij - μ - di - for any observation in the two way table of additive effects of components of variability and environments respectively.

For the joint regression analysis the expected mean sum of squares are:

Items	df	MS
Lines (difference between genotypes)	(t – 1)	s Σ_i (di)^2 / (t – 1)
Environments (joint Regression)	(s – 1)	t Σ_j (ε_j)^2 / (s –1)
Interaction		
Line X Environments	(t–1) (s–1)	Σ_i (β_i)^2Σ_j(ε_j)^2/(t-1)
Heterogeneity Between regression	(t – 1)	
Remainder (TSS – others)	(t–1) (s–2)	Σ_{ij} δ^{2}ij/(t-1)(s-12

Errors: Error MS of the initial anova for the 20 lines over 6 environments

Where 's' represents the number of environments and 't' as the number of inbred lines.

4. Results and Discussion

Table 1
Analysis of Variance

Character	df	Mean sum of squares		Error
		Variety	Environment	
1. Appearance of 1st trifoliate leaves (d)	5	10.1391	912.4561*	3.253617
2. Appearance of 1st green pod (d)	5	83.2912*	2135.5624*	1.255906
3. 50% flowering time (d)	4	31.4084	834.7900	840022.18
4. Appearance of 1st mature pod (d)	4	69.5566*	1547.8666*	2.471346
5. Length of leaf (cm)	5	0.1006	22.6883*	0.123741
6. Breadth of leaf (cm)	5	0.1403	10.6901*	0.071112
7. Green leaves/pod (no)	5	4469.45	983797.71*	1942.3413
8. Pl. height at 50% flowering (cm)	5	80.7871	45801.3253*	135.397665
9. Length of pod)cm)	4	1.4190	12.3113*	1.423972
10. Breadth of pod (cm)	4	0.0030	0.16689*	0.002908
11. Open stomata/mfa (no)	5	7.5227	108.3738*	7.261190
12. Close stomata/mfa (no)	5	1.3781	2.1482	0.710339
13. Internode length (cm)	5	0.5923	37.2867*	0.379440
14. Dry matter of single plant (g)	4	71.4666	28982.49*	46.177780
15. Pods/plant (no)	4	1.6031	50.3839*	0.892352
16. Seed/plant (g)	4	3.9076	23.8218	1641.3432
17. Seed/pod (g)	4	5.6987*	32.1323*	1.148102
18. Nodules/plant (no)	5	2482.44*	15279.1283*	272.8476
19. Seeds/pod (no)	4	0.7706	3.46019*	0.143196
20. 100 seed (g)	4	0.9908	0.716658	0.403067

* indicates significant at P 0.05 level

Considering that all the varieties tested here were uniformed genetically, the expected mean sum of square for error would be purely a random environmental variance. The mean sum of squares between varieties would consist (i) the genotypic variance and (ii) the environmental variance. Thus the expected mean sum of squares would be:

$E(MS_v) = d^2 e + \ d^2 g$
$E(MS_e) = d^2 e$ and therefore, $d^2 g = MS_v - MS_e / r$ where, r = number of replications.
Thus, phenotypic variance $(d^2 p) = d^2 g + d^2 e$; and
h^2 (heritability in broad sense) = $d^2 g / d^2 p$

The components of variances of all characters were calculated following the expressions as above.

Table 2
Joint regression analysis of inbred lines
(Analysis of Variance)

Character	Mean Sum of Squares				
	Lines (df=19)	Environments (df = 4 or 5)	Heterogeneity between Regression (df = 19)	Remainder (df = 57 or 76)	Error (df=76 or 95)
1	10.1392*	912.4561*	12.4491*	0.9547	3.0536
2	83.2913*	2135.5624*	5.7337*	0.1365	1.2559
3	31.4084	834.7900	4.6371	1120020.5830	840022.1874
4	69.5450*	1547.8343*	6.2730*	0.9074	2.4713
5	0.1008	22.6883*	0.1854*	0.1546	0.1237
6	0.1403*	10.8304*	0.2642*	0.0228	0.0711
7	4469.4456*	988267.1591*	27.533.4192*	350.1640	1942.3413
8	80.7871	45801.3253*	275.8664*	100.2804	135.3976
9	1.4190	12.3113*	1.2619	1.4779	1.4239
10	0.0030	0.1699*	0.0014	0.0033	0.0029
11	139.7572*	132.2345*	23.8616*	4.9263	7.2611
12	1.3781	2.1482*	0.0618	0.8724	0.7103
13	0.5923*	37.2867*	0.5345	0.3406	0.3794
14	71.4666*	28.9824*	121.8055*	20.9685	46.1777
15	1.6031*	50.3839*	1.1596	0.8032	0.8923
16	3.9076	23.8218	1.9249	2187.8160	1641.3432
17	5.6987*	32.1323*	1.3633	1.0763	1.1481
18	2484.4434*	15279.1283*	502.3479*	215.4725	272.8476
19	0.7706*	2.7681*	0.1442	0.1428	0.1431
20	0.9908*	0.7166	0.3271	0.4283	0.4030

MSS compared either by remainder MSS or error MSS at 5% level

Out of the 20 characters studied, only 4 characters were found to be showing significant variation and thus the varieties significantly varied between each other for these characters. All these 4 characters also carried a significant variation between the environments. Such results were expected to show higher genetic variances and consequently higher heritability.

The data were subjected to single line regression for the 20 characters followed by joint regression analysis. 20 lines X 20 characters i.e. 40 single line regressions were estimated and 400 regression coefficient values (b_i) were available to understand stability of each line for each character. The joint regression analyses have been shown in Table 2.

In view of the urgent need to enhance production of protein as well as medicinal crops for the development health status of the society in general the pulses are given the primary importance which constitutes the major sources of high nutritional food for human consumption.

20 X 20 characters yielding 400 single regressions with 400 regression coefficients values (b_i) were available to understand the organization of stability. This regression analyses help us to find out the most suitable genotype(s) which are capable to produce highest yield of plant canopy for the commercial endeavour. Thereafter, on assumption of the genotypic behavior the most suitable genotypes were selected for further diallele breeding programme. The data of F_1 lines have been cited in another paper of the authors.

5. Conclusion

Considering all these characteristics of the legume it can be recommended that the negative additive genes should be accumulated for evolving stable lines against the adverse environmental conditions. Along with this positive (b_i) values near around 0.00 are also desired since they will provide more flexibility that would generate high GEI and give scope of selection against the same environment.

REFERENCES

Aubery M. and Bourrillon R. 1973. Changes in chick embryonic cell growth induced by a specific lectin. *IRCS libr. Compend* 1(7): 1-7.

Baluch M. A. A. and Soomro M. P. M. 1968. Correlation studies in gram. *West Pakistan J. Agric. Res.*, VI (3): 26-32.

Bhaumik P. K. 1971. Genetical Studies in *Dolichos biflorus L.* PhD Thesis, Calcutta University, July, 1971.

Bucio-Alanis L. 1966. Environmental and genotype-environmental components of variability. I. Inbred Lines. *Heredity*, 21: 387-397.

Chatterjee S. D. 1970. Studies on application of discriminate functions in classification and selection problems in horsegram (*Dolichos biflorus* L.), M.Sc. (Ag.) in Agricultural Botany Thesis, Calcutta University, 1970.

Gupta V. P. and Ramanujam S. 1974a. Genetic artitecture of yield and its components in Bengal gram. *Indian J. Gent.* XXXIV (A): 793-799.

Kulkarni R. N. and Shivasankan G. 1978. Mutagenic effect of gamma rays and ethyl methanesulphonate in horsegram (*Dolichos biflorus L.*). *Genetica Agraria,* 32 (1/2): 65-71.

Mahapatra I., Mahindra C., Singh Dayanand and Singh R. N. 1975. Pulses in cropping systems. *Ind. J. Genet. Pl. Breed.* 35(2): 188-193.

Carter W.G. and Etzler M. E. 1975. Isolation and characterization of cyanogens bromide fragments and a glycopeptide from the *Dolichos biflorus L.* lectin. *Biochem* 14: 5118-5122.

Pere M., Font J. and Bourrillon R. 1974. moleculaer weight and submit structure of the lectin of *Dolichos biflorus L.* Biochem, *BioPhysa Acta.* 365(1): 40-46.

Perkins Jean M. and Jinks J.L. 1968a. Environmental and genotype-environmental components of variability. III. Multiple Lines and crosses, *Heredity.* 28: 51-70.

Perkins Jean M. and Jinks J.L. 1968a. Environmental and genotype-environmental components of variability. IV. Non-linear interaction for multiple inbred lines. *Heredity.* 23: 525-535.

Perkins Jean M. and Jinks J.L. 1969. Environmental and genotype-environmental components of variability. III. Multiple lines and crosses. *Heredity.* 23: 339-356.

Perkins Jean M. and Jinks J.L. 1973. The assessment and specificity of environmental and genotype-environmental components of variability. *Heredity,* 30: 111-126.

Pooni H. S. and Jink J. L. 1985. Predicting the properties of first cycle inbred and second cycle hybrids extractable from two, three and four parent crosses. *Heredity.* 32: 167-192.

Prakash D. and Purohit R. M. 1972. Chemical examination of the fixed oil from the seeds of *Dolichos biflorus L. Ind. Oil Soap J.* 37(II): 284-288.

Sen S. N., Sengupta K. and Mukherjee D. K. 1965. Adaptability of certain selection of gram in West Bengal. *Ind. Agric.* IX(2): 141-150.

Singh S. P. and Mehendiratta P. D. 1970. Path analysis and selection indices for cowpea. *Indian J. Genet.* XXX: 471-475.

Singh M. P. and Mehra R. B. 1980. adaptability studies in Bengaal gram (Cicer arietinum L.) *Trop. Grain. Leg.* Bull. XIX: 51-54.

Tah J. and Dasgupta P. 1988. Variety X Environment interaction and its contributing components in Indian minor pulse (Dolichos biflorus L). *Genome.* 30: Suppl. 1. XVIth International Cong. Gent. Proc. 33.14.48.

Talbot F. Craig and Etzler E. Marilynn. 1978. Development and distribution of *Dolichos biflorus L.* Lectin as measured by radioimmuno assay. *Plant Physiol.* 61: 847-850.

Tomar G. S., Singh L., Sharma D. and Deodhar A. D. 1973. Phenotypic stability of yield and some seed characteristics in West Bengal gram (*Cicer arietinum* L.) varieties JNKVV. *Res. J.* VII: 35-39.

8

Phyto-Ecological Diversity of Some Less-known Wild Plants With Special Reference to Ethnobotany of Bodo-Kachari People of Rangia Sub-Division, Kamrup, Assam

S. Sarkar and N. J. Das

1. Introduction

Being the gateway of the whole of Northeast India, Kamrup district is characterized by highly humid sub-tropical climate. The district embraces evergreen and semi-evergreen forests, grasslands including riparian forests and swamps. Each has its own types of vegetation and ecological characteristics. Geographically, Kamrup district is also important in finding its position in between the geographically important landmasses and having the valley of deposition by the Brahmaputra river and its several tributaries.

Rangia sub-division is situated in the northern part of Kamrup district. It is bounded on the north by Nalbari district, on the south by Hajo and Guwahati, on the east by Darrang district and on the west by Bhutan and Nalbari district. The sub-division does not differ much with the rest of Kamrup district, which is a part of Brahmaputra valley, the major physiographic unit of Assam. The climatic condition of the area does not differ much from that of Assam valley. The area has a sub-tropical climate with warm summer and dry cool winter. The soil of the area consists of greater percentage of sand. Along the Baralia River, soil ranges from loamy sand to silt loam. The sub-division comprises of elongated hills of various altitudes and recent alluvial deposits of the Brahmaputra River. The sub-division is divisible into three regions, viz. (i) the hilly part, (ii) the alluvial tract and (iii) the swampy areas.

The region is inhabited mostly by Bodo-Kachari people. The people, traditionally living close to nature, depend much on forests and utilizing many of the plant resources for their food, clothing, shelter, cottage industries, medicines etc. The denudation of forests has not only adversely affected the ecological balance but has also affected the economy of the aboriginal people of the district. Systematic and thorough investigations are of paramount importance to record the occurrence of plant species and their proper identification. So, in order to assess the botanical potentialities for useful exploitation, researches on economic utilization of these plants are therefore considered the most essential before they may be lost forever.

To undertake such investigation, it is necessary that floristic and economic diversity of the region should be well known and only when all the plants are properly collected and preserved, accurately identified and described for a system of classification, then it would be possible to assess the vegetable wealth of the region.

2. Methods of Present Study

Collections have been made during the period 2004-05 for almost all the seasons of the year so as to collect maximum number of herbaceous specimens from the area. Emphasis has been made to collect plant materials preferably in its flowering/fruiting stage from the natural habitat. Thorough observations and recorded field data have been made on the spot of collection of individual plant species with regards to habit, habitat, ecological adaptation including the associated plants, distribution pattern, nature of the roots or rhizomes/ bulbs etc. The specimens are preserved dried or in preservatives. As most of the specimens are small herbs, they were uprooted and collected entirely.

Information regarding the vernacular name of the individual plant species and their utilization by the local inhabitants have been collected and recorded on the spot. In several occasions, the local people including the headman, women folks or young ones accompanied to the study area in searching for the useful plants. They provided useful information on the utility of these plants in their day to day life.

3. Discussions on the Results

The study was made during 2004-2005. An attempt has been made to make an inventory of some less-known wild plants with special reference to diversity of floristic, ecological and ethno botany of Bodo-Kachari people of Rangia Sub-division in Kamrup district of Assam. Observations in the field and literature studies have enabled to record the number of wild plant species available with their economic importance in the region. Apart from these, the impact of diverse plant resources in socio-cultural life of the sub-division has also been studied.

During the field survey, different aspects of diversity of economic utilization of plants have been studied and found that the area is rich in wild herbaceous plants. The sub-division is also rich in forest wealth. The forest resources including food and vegetable plants, timber, medicinal plants, food adjuncts, beverages and plants of miscellaneous uses. The species enumerated have been categorized based on the mode of utilization of plant parts. The study revealed that the species used by the ethnic people are mostly wild herbs. Out of 32 families dealt with, 8 are monocots and 24 are dicots i.e. the ratio of monocots to dicots is 1:3.

The region supports luxuriant growth of floristic elements among which herbaceous plant species account for a major bulk. The vegetation of the surveyed area can be broadly categorized in four main ecosystems viz.

1. Forest Ecosystem.
2. Wasteland Ecosystem.
3. Aquatic Ecosystem.
4. Agricultural Ecosystem.

Ethno botanical information has been recorded from the tribe of Bodo-Kachari people. It has been observed that these tribal people have their own unique culture, tradition, medicinal practices, ritual belief etc. and they make use of various wild plants which are mostly herbaceous in meeting their day to day requirements.

The common pattern seen in the mode of living of the Bodo-Kachari people in villages is that they normally put barriers all around their homestead. This ensures privacy and aesthetic beauty to a residential house. The economy of ethnic group of people of the area is basically agricultural. The methods and quality of their agriculture differ specially in between hills and plains, depending on the nature of topography and the level of development in the society. In the plains there are areas where modern methods are followed but in the study area the traditional methods are still followed.

The Deuri, an assistant to the village priest (Ojai) usually conduct the atonement proceedings. He is elected on the general consensus of all on the basis of his character and ability to guide socio-religious rites.

The Sizu plant *(Euphorbia splendens)* is considered to represent the Bathou, the supreme deity of adoration. Therefore, every traditional Bodo-Kachari household is seen planting a Sizu tree along with a sapling of Jatrası (*Justica gendarussa*) and tulsi (*Ocimum sanctum*) in the northern corner of their courtyard.

Kherai, an annual religious community festival of the tribe is another occasion when the entire atmosphere is agog with much enthusiasm and festive fervor. To some, it is a national festival of the community.

Nature has much influence in their traditional mode of living and even in their religious belief. The Bathou puja is held at the Second Tuesday of January. The Sizu plant is representing the Bathou, the supreme deity of adoration. Therefore, every traditional Bodo-Kachari household is seen planting a Sizu tree in the Northeast side of the courtyard. The kitchen is situated in the northern side and the rice storage is situated in front of Sizu plant. The Sizu plant is surrounded by 18 bamboo sticks, which means 18 gods and goddesses. One pair of bamboo sticks means one male and one female. The 18 sticks are surrounded by five pairs of long bamboo sticks in alternate way. Near the Sizu plant, 18 pairs of grasses (Batah in Bodo) are kept which means 18 gurus according to their traditional belief. These 18 gurus teach them their mode of living in their daily life.

Some of the important species are enumerated below

Sl. No.	Botanical name & family	Local Name in Bodo	Plant part(s) used	Mode of use
1.	*Allium sativum* L. (Liliaceae)	Samrum	Leaves	Used in burn injuries
2.	*Centella asiatica* L. (Apiaceae)	Manimuni	Leaves	Eaten to cure dysentery
3.	*Cissampelos pareira* (Menispermaceae)	Dudlata	Stem	The juice of stem is used to reduce the swelling of hands and legs
4.	*Commelina benghalensis* (Commelinaceae)	Nasraiphulanti	Stem & leaves	Both are used to stop bleeding
5.	*Desmodium gangeticum* L. (Papilionaceae)	Saptimah	Leaves	Used to reduce ulcer
6.	*Evolvulus nummeroloides* (Convolvulaceae)	Singrimukhi	Stem & leaves	Whole plant is used to reduce tonsillitis
7.	*Houttuynia cordata* Thunb (Soururaceae)	Mesendari	Leaves	Used to reduce body ache

4. Conclusion

From the analysis it was observed that the herbaceous plant species are of economic importance for the benefit of the local people. Most of the herbaceous species thrive well under diffused sunlight and more particularly species with bulbs, rhizomes and corms can't grow well in open area. Deforestation and monoculture of economically important species result in the scarcity of the population of many naturally occurring species. Some of them are found to be endangered and gradually facing extinction. *Costus speciousus, Acorus calamus, Dioscoria bulbifera* etc. are now scarcely found in the sub-division. Unless proper preservation and protection measures are taken and people are educated properly, genetic diversity of plants in the region and their resources will be lost forever.

REFERENCES

Babu C.R. 1997. *Herbaceous flora of Dehradun*, New Delhi.

Bordoloi B.N. et al. 1987. *Tribes of Assam*, Part – I, Govt. of Assam.

Deb D.B. 1981 and 1983. *Flora of Tripura State*, Vols. 1- 2, BSI, Howrah, Today and Tomorrow's Publishers, New Delhi.

Hooker J.D. 1872 – 1887. *Flora of British India*, Vols. 1 – 7, London.

Jain S.K 1990. *Observation of Ethnobotany of the tribals of Central India, Contribution of Ethnobotany of India*, pp. 59 – 73.

Kanjilal U.N. et al. 1934 – 40. *Flora of Assam*, Vols. 1-4, Govt. of Assam, Shillong.

Kar S. K. and Panighari 1963. The Rubiaceae in Assam and NEFA, *Bull. Bot. Serv. India*, Calcutta (227 – 237).

Rao A.S. 1977. Floristic studies in North-East India (Old Assam Region), *Bull. Bot. Surv. India*, 19 (1 – 4) 56 – 60.

Rao A.S. and Rabha L.C. 1966. Contribution to the botany of Kamrup district (Southern part), Assam, *Bull. Bot. Surv. India*, 8: 296 – 303.

Sarkar S. *et al.* 1989. A study on the utilization of indigenous plants of Karbi Anglong district of Assam – II, Food and vegetables, medicinal plants, *J. Ass. Sci. Soc*, 31 (2): 43 – 54.

Takhtajan A. 1969. *Flowering Plants: Origin and Dispersal* (transl. By C. Jerreri), Edinburgh.

Watt G. 1972. *A Dictionary of Economic Production of India*. Vols. 1-6, Calcutta.

9

Proctective Role of *Hibiscus rosa-sinensis* L. Leaf Extract Against Piroxicam-induced Hepatotoxicity in Mice

C.R. Sahu and J. Ghosal

1. Introduction

Drugs can induce oxidative stress by generating free radicals that are mostly available as by products or as an aerobic metabolic product. These free radicals when generated excessively at cellular level may cause damage to tissue proteins, nucleic acids, membrane lipids and have association with many age-related problems (Halliwell & Gutterridge, 1994; Wang & Jiao, 2000). The balance between the production and scavenging of reactive oxygen species (ROS) or free radicals determines the susceptibility of the body to oxidative damage. Though the self-antioxidant defense mechanisms of an organism minimize the production of free radicals, thus protecting the oxidative damage, but may not be sufficient to prevent the damage entirely. The level of these defense mechanisms may be altered through the introduction of the drugs and there is ineffective scavenging of free radicals which may cause tissue injury.

Non-steroidal anti-inflammatory drugs (NSAIDs) are important therapeutic class of drugs used to suppress pain and inflammation in case of rheumatoid arthritis, osteoarthritis and other inflammatory diseases (Abramson & Weissman, 1988). Besides being anti-inflammatory, these drugs are analgesic and anti-pyretic, often used for the relief of non-specific fever conditions (Gettigan & Hengy, 2000). More than 100 million NSAIDs are prescribed throughout the world (Steinmeyer, 2000). Piroxicam (an acidic carboxamide) which belongs to a chemical subgroup of NSAIDs that are oxicam derivatives, a class of enolic acids are advocated for use in various painful and inflammatory conditions, specially as single largest group of NSAIDs associated with the palliation of symptoms rheumatoid arthritis, ankylosing spondylitis and musculoskeletal disorders (Roberts & Morrow, 2001). Recently, piroxicam (a non-specific COX inhibitor) has also gained attention as an effective therapy for tumors, colorectal and invasive bladder cancers (Ronald, 2002). Despite its widespread use, it can cause many adverse effects such as severe gastrointestinal toxicity, ulcerogenic gastropathy, renal homeostatic abnormalities (Murray & Brater, 1993), proteoglycan synthesis from chondrocytes, foetotoxicity and other processes depending of prostaglandins (Brune & Linder, 1992). Like other NSAIDs, the mechanism of action of piroxicam involves reduction of prostaglandin synthesis by inhabiting cyclooxygenase (COX) enzyme through competitive antagonism for arachidonic acid (Vane *et al.* 1998). Accordingly, by inhibiting the prostaglandin synthesis, indirectly reducing gastro-protective mucin secretion and an increased risk of ulceration arises. The central roles of liver in drug metabolism predispose them to toxic injury. Since piroxicam is metabolized in liver, there is a possibility of injury in the liver. The toxicity develops due to piroxicam is mediated through oxidative stress, that lead to lipid peroxidation and free radical generation. Accordingly there occurs hepatic dysfunction and failure.

With this view, much attention has been paid on the protective effect of some naturally occurring antioxidants on the living system. In ayurvedic treatment, different parts of the plant have been prescribed for different ailments. The active principles from the plants have been identified and become useful in curing various diseases. The present investigation was undertaken to study the protective effect of the alcoholic leaf extract of *Hibuscus rosa-sinensis* (AEH), on antioxidant status against piroxicam induced hapatotoxicity in mice.

2. Materials and Methods

Plant Material

Hibiscus rosa-sinensis (Malvaceae) was identified by plant taxonomist of Botany Department, Kalyani University. The matured fresh green leaves were collected, shed, dried and powdered (about 500 gms) and later subjected to extraction with 70% ethanol (1.5 liter), then made them a semisolid mass under reduced pressure following the methods described by Srinivasan *et al.* 2001 and Essa *et. al.* 2005. The extract was dissolved in double distilled sterile water and was used in the investigation.

Experimental Animals

Swiss albino male mice (*Mus musculus*) (20-25g) were purchased from the supplier and were acclimatized in the laboratory for 7 days. They were maintained and housed in polypropylene cages in the departmental animal house and provided with standard pellets and water *ad libitum*. All experiments were approved by the ethical committee (vide No. 892/ac/05) constituted through CPESCA. The chemicals used were of AR Grade.

Experimental Design

The mice were selected randomly and divided into four groups of ten each and fed with normal diet. The extract, the dissolved drug and double distilled sterile water were administered orally.

Group A: Control (only normal diet)

Group B: Mice treated with AEH (30 mg kg^{-1} b.w.) for 15 days.

Group C: Mice treated with piroxicam (6.6 mg kg^{-1} b.w.) for 15 days.

Group D: Mice treated with piroxicam and AEH for 15 days.

Hepatoprotectivity

At the end of the scheduled treatment the blood samples were collected from sacrificed mice by cardiac puncture under ether anesthesia and allowed to clot at room temperature for 45 minutes. Serum was separated by centrifugation at 4,000 rpm at 4°C for 15 minutes and was used for the assay of serum marker enzymes e.g. aspartate amiontransferase (AST), alanine aminotransferase (ALT), following the method of Reitman & Frankel (1957) and alkaline phosphatase (ALP) was determined by the method of Kind & Kings (1954). It was well known that activities of serum transaminases and phosphatases generally represented the functional status of the liver. Liver samples were taken and immediately washed in ice-cold saline for removal of blood as much as possible. It was weighed and 10% (W/V) tissue homogenate was prepared in phosphate buffered saline (pH 7.2) to measure lipid peroxidation (LPO) in terms of TBARS (Thiobarbituric acid reactive substances) following the method of Neihaus & Samuelsson (1968). GSH (Reduced glutathione) and SOD (Superoxide Dismutase) were measured by the method of Ellman (1959) and Kakkar *et al.* (1984) respectively. However, the other two hepatoprotective enzymes CAT (Catalase) and GPx (Glutathione peroxidase) were measured as per the methods of Sinha (1972) and Rotruck *et al.* (1973).

Histopathology

For histopathological studies, a portion of the liver was taken and fixed in 10% formalin. After scheduled period of fixation and dehydration, routine histological technique was followed for section cutting at 6ìm. The tissue sections were thoroughly stained with haematoxylin and eosin stains and observed under microscope and good stained sections were photographed (Luna, 1968).

Statistical Analysis

The values were represented as mean ± SD of five repeated determinations and statistical analysis was carried out by analysis of variance (ANOVA). Difference was considered to be Significant when P< 0.05.

3. Results

Table 1 shows the activity of serum enzymes in normal and experimental groups. The enzymes activity was significantly higher in piroxicam treated mice. Mice co-administered with piroxicam and AEH showed significantly lower activity when compared to corresponding piroxicam treated group. Mice treated with AEH alone did not alter the enzyme activity when compared to the normal values.

Table 1

Effect of AEH on changes on serum marker enzymes of normal and treated mice

Group	ALT*	AST*	ALP*
Normal	10.26 ± 0.02^{a}	49.88 ± 1.17^{a}	72.73 ± 5.94^{a}
AEH (30 mg kg^{-1})	10.60 ± 0.98^{a}	48.74 ± 2.04^{a}	71.40 ± 5.57^{a}
Piroxicam (6.6 mg kg^{-1})	16.78 ± 0.53^{b}	62.20 ± 1.96^{b}	81.10 ± 3.68^{b}
Piroxicam (6.6 mg kg^{-1}) + AEH (30 mg kg^{-1})	12.87 ± 0.77^{c}	51.82 ± 2.32^{c}	73.44 ± 4.32^{c}

Values are mean ± SE (n = 10)
Values not sharing a common superscript (differ significantly at P < 0.01
* Activities are expressed as Units/ml
ALT-Alanine aminotransferase, AST-Aspartate aminotransferase, ALP -Alkaline Phosphatase

It has been observed from Table 2, that the level of LPO was higher, whereas, the levels of SOD, CAT, GSH-Px and GSH were significantly low in the liver of piroxicam treated mice. Mice treated with piroxicam and AEH showed significantly (P<0.05) low levels of LPO and significantly (P<0.05) elevated levels of SOD, CAT, GSH- Px and GSH when compared with corresponding piroxicam treated group.

Table 2

Groups	SOD U/mg Protein	CAT U/mg Protein	GSH-Px µmol/g	GSH µmol/g	LPO (TBARS) mM MDA/100g
Normal	13.73 ± 0.27[a]	8.42 ± 0.40[a]	37.46 ± 1.08[a]	3.79 ± 1.08[a]	1.62 ± 0.08[a]
AEH (30 mg kg^{-1})	13.64 ± 0.32	8.46 ± 0.41	37.01 ± 5.23	3.76 ± 1.21	1.60 ± 0.12
Piroxicam (6.6 mg kg^{-1})	7.52 ± 0.45[b]	5.86 ± 0.61[b]	24.87 ± 4.20[b]	2.02 ± 0.64[b]	2.42 ± 0.05[b]
Piroxicam (6.6 mg kg^{-1}) + AEH (30 mg kg^{-1})	12.35 ± 0.38[c]	8.34 ± 0.33[c]	35.89 ± 5.06[c]	3.53 ± 1.13[c]	1.69 ± 0.14[c]

Values are mean ± SE (n=10). Values not sharing a common superscript (a, b, c) differ significantly at $P < 0.05$)

SOD-Superoxide dismutase, CAT-Catalase, GSH Px- glutathione peroxidase, GSH-Reduces glutathione, LPO-Lipid peroxidation, TBARS-Thiobarbituric acid reactive substances MDA- Malondialdehyde

Histopathological Observation

Liver sections of the control mice showed normal histology at the centrilobular and periportal regions (Fig. 1). No significant alternations were observed only in AEH treated mice. Histopathological observation of piroxicam treated liver showed some abnormalities compared to the tissue sections of the control / normal liver. Some of the abnormalities encountered as fatty degeneration, necrosis and sinusoidal dilations (Figs. 2, 3, 4). Administration of the AEH concurrently with the drug helps to maintain the normal architecture of the liver except few mild irregularities (Fig. 5).

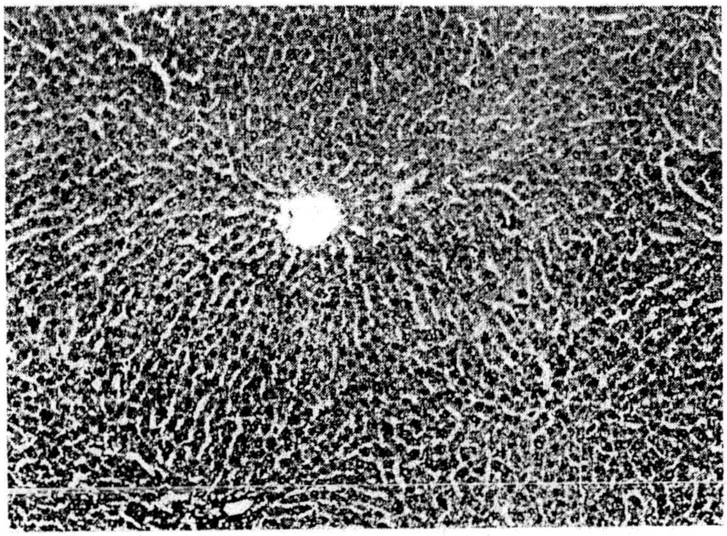

Figure 1. T.S. of Liver of control mice showing normal histological architecture (H & E, 200X)

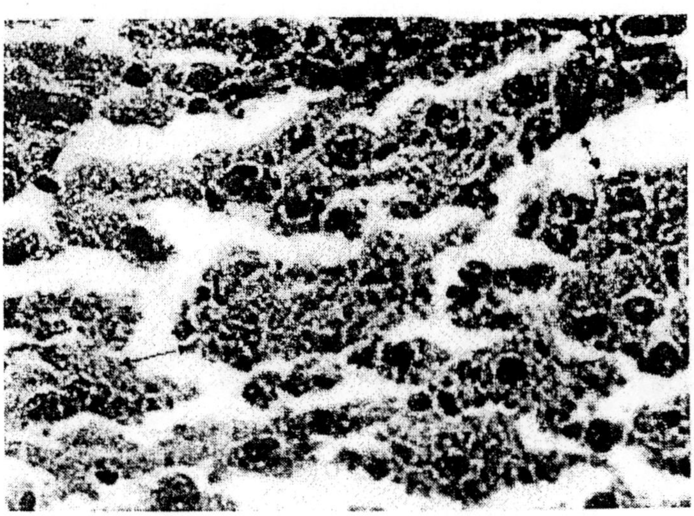

Figure 2. T.S. of Liver of piroxicam treated mice showing sinusoidal dilation (H & E, 400X)

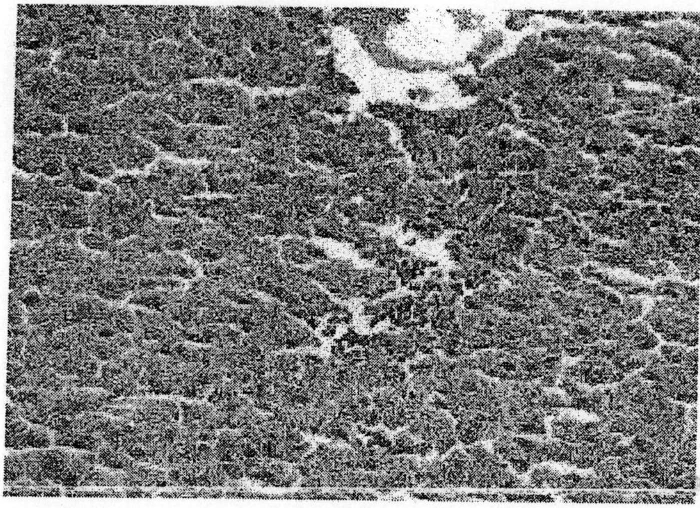

Figure 3. T.S. of Liver of piroxicam treated mice showing necrosis with focal inflammatory reaction (H & E, 200X)

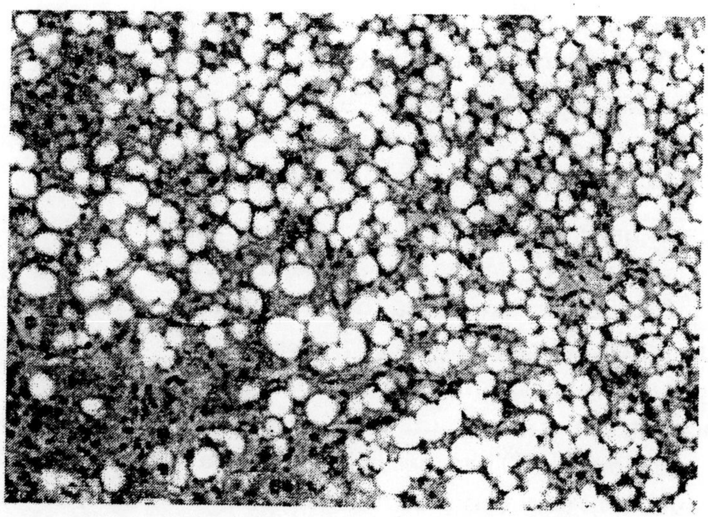

Figure 4. T.S. of Liver of piroxicam treated mice showing fatty changes (H & E, 200X)

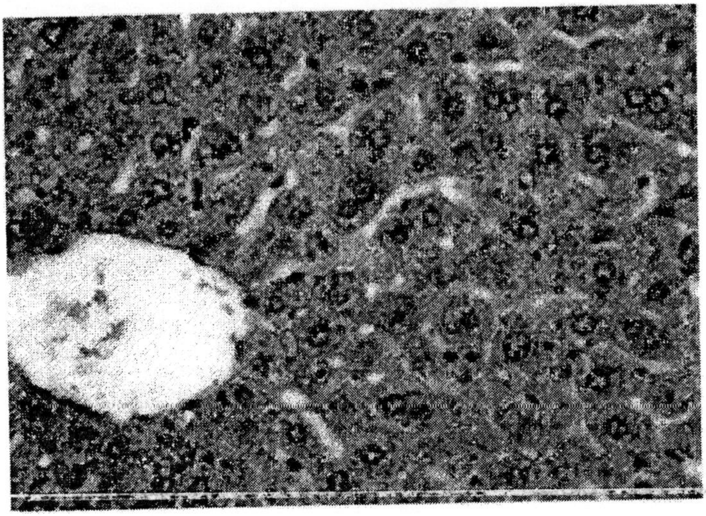

Figure 5. T.S. of Liver showing normal arrangement of cord cells in piroxicam and AEH co-administered mice (H & E, 400X)

4. Discussion

It is a well established fact that the oxidative stress developed due to the introduction of the xenibiotics causing damage to the cells via oxidative stress-mediated lipid peroxidation (Galati *et al.* 2002; Usoh *et al.* 2005). The use of various dietary antioxidant treatment in terminating or reducing free radical attacks (Neff, 1997; Chung *et al.* 1999; Baubles *et al.* 2000), that are involved in various diseases are also an important part of the antioxidant mechanism. The antioxidant may act as free radical scavengers, reducing agents and activators of antioxidative defense enzymes system to suppress the radical damage in biological system (Zielinski & Kozlowska, 2000).

The piroxicam is most commonly used drug causing hepatotoxicity in the experimental study. The peroxidative degradation of the lipid membrane is one of the principal causes of hepatotoxicity. Studies have revealed that the mechanism of piroxicam hepatotoxicity relates both to impairment of adenosine triphosphate (ATP) synthesis by mitochondria and production of active metabolites, particularly 5-hydroxy piroxicam, which causes direct cytotoxicity (Rabinivitz *et al.* 1990). Induction of mitochondrial membrane permeability transition (MPT) has also been shown to be important in NSAID- induced liver injury, resulting in generation of reactive oxygen species, mitochondrial swelling and oxidation of nicotinamide adenine dinucleotide phosphate (NADP) and protein thiols. Similar events might also occur in this case. The present study shows that the levels of serum marker enzymes, e.g. AST, ALT and ALP in experimental Group C mice become elevated, which may be due to the liver damage caused by drug induced free radical generation. The administration of AEH in Group D has significantly reduced these liver enzyme levels.

The present experiment also shows the elevated levels of lipid peroxidation in piroxicam treated (Group C) animals. The increase in TBARS levels in liver suggests the enhance lipid peroxidation generating free radical which is deleterious for the cell membrane. Increased lipid peroxidation damage the membrane function considerably by decreasing membrane fluidity and changing the activities of membrane bound enzymes leading to oxidative stress. This phenomenon also suggests the failure of antioxidant defense mechanism to some extent. Treatments with AEH (Group D) significantly prevent these changes by suppressing lipid peroxidation level. This may be due to the free radical scavenging properties of AEH (Amin &Hamza, 2005; Liu *et al.* 2002). Since, AEH in Group D animals has significantly increased the SOC, CAT, GSH and GSH-Px contents of the liver, it may also be important in preventing hepatotoxicity caused by the drug. The drug decreased the antioxidant level in group C, whereas AEH treated group (Group B) is almost similar to the normal group (Group A).

During cellular metabolism, reactive oxygen species (ROS) are formed continuously which are normally prevented or scavenged by a host of antioxidants (Datta *et al.* 2000; Halliwell & Gutteridge, 1999). The SOD and CAT may play an important role in detoxification of superoxide anion and hydrogen peroxide respectively and thus protecting the ROS-induced damage. GSH in conjunction with GSH-Px helps protection against free radicals and the toxic compounds (Cotgreave & Gardes, 1998; Rahman & Macnee, 2000). The increased level of liver antioxidant enzyme activities in piroxicam and AEH treated mice may be due to the presence of chemical compounds notably flavonoids in AEH which may have a positive role in reducing the oxidative stress by inducing cellular antioxidant enzymes.

Histopathological studies showed that the drug induces fatty degeneration and necrosis in the liver tissue. Treatment of AEH shows the reversibility of the original condition in liver tissue thus indicating the protection against drug-induced liver toxicity. Further studies are needed to investigate and isolate the active ingredients for possible mechanism action of the extract in controlling toxicity.

REFERENCES

Abramson S.B. and Weissman G. 1988. The mechanism of action of Non-steroidal anti-inflammatory drugs. *Arth. Rheum.* 32:1.

Amin A. and Hamza A.A. 2005. Hepatoprotective effects of Hibiscus, Rosmarinus and Salvia on azathioprine-induced toxicity in rats. *Life Sciences*, in press.

Baubles A., Decker E.A. and Dale F.M.C. 2000. An antioxidant effects of aqueous extracts from wheat-based ready to eat breakfast cereals. *Food Chem* 68:1-6.

Brune K. and Linder J. 1992. Side effects of anti-inflammatory drugs. *Inflammation and Drug Therapy Series*. 5: 198-203.

Chung H.S., Chang L.C., Lee S.K., Shamon L.A. and Kinghorn A.D. 1999. Flavonoids constituents of caffeic acid and its related hydrocinnamic acid compounds. *J Agri Food Chem* 45: 36-41.

Cotgreave I.A. and Gerdes R.C. 1998. Recent trends in glutathione biochemistry-glutathione protein interaction: a molecular link between oxidative stress and proliferation. *Bichem Biophys Res Commun* 242: 1-9.

Datta K., Sinha S. and Chattopadhy P. 2000. Reactive oxygen species in health and disease. *The Natl Med J India* 13: 304-310.

Ellman G.L. 1959. Tissue sulfahydryl group. *Ach Biochem Biophys*. 82: 70-72.

Essa M.M., Subramanian P., Suthakar G., Manivasagam T. and Dakshayani K.B. 2005. Protective influence of *Pongamia pinnata* (Karanja) on blood ammonia and urea levels in ammonium chloride-induced hyperammonemia. *J Applied Biomedicine* 3: 13-24.

Galati G., Tafazoli S., Sabzevari O., Chan T.S. and O'Brien P.J. 2002. Idiosyncratic NSAID drug induced oxidative stress. *Chem Biol Interac* 142: 25-41.

Haliwell B. 1994. Antioxidant in human health and disease. *Ann Rev Nutr* 16: 33-50.

Haliwell B. and Gutterridge J.M.C. 1999. Oxidative stress: adaptation, damage, repair and death. *In Free Radicals in Biology and Medicine*, eds. Oxford University Press. Oxford. U.K. pp. 284-330.

Kakkar P., Dass B. and Visvanathan P.N. 1984. A modified spectrophotometric assay of superoxide dismutase. *Indian J Biochem* 197: 588-590.

Kind P.R.N. and Kings E. 1954. Estimation of plasma phosphatase by determination of hydrolysed phenol with antipyrine. J Clin Pathol 7: 322-330.

Liu C., Wang J., Chu C., Cheng M. and Tseng T. 2002. In vivo protective effect of protocatechuic acid on tert-butyl hydroperoxide-induced rat hepatotoxicity. *Food Chem Toxicol* 40: 635-641.

Luna L.G. 1968. *Manual of Histological Staining*. New York: McGraw Hill Company.

Murray M.D. and Brater D.C. 1993. Renal toxicity of Non-steroidal anti-inflammatory drugs. *Annu Rev Pharmacol Toxicol*. 33: 435-465.

Neff J. 1997. Big companies take nutraceuticals to heart. *Food Processing* 58: 37-42.

Neihaus W.G. and Samuelsson B. 1968. Formation of malondialdehyde from phospholipids arachidonate during mitochondrial lipid peroxidation. *Eur J Biochem* 6: 126-130.

Rabinivitz M. and VanThiel D.H. 1990. Hapatotoxicity of non-steroidal anti-inflammatory drugs. *Arth. Rheum.* 10: 1449-1461.

Rahman I. and Macnee W. 2000. Oxidative stress and regulation of glutathione in lung inflammation. *Eur Regir J.* 16: 534-554.

Reitman S. and Frankel S.A. 1957. Colorimetric method for the determination of glutamic oxaloacetic and glutamic pyruvic transminase. *Am. J. Clin. Pathol.* 8: 56-63.

Roberts L.J. and Morrow J.D. 2001. Analgesic, antipyretic and anti-inflammatory agents and drugs employed in the treatment of gout. In Goodman L.S. and Gilman A. eds. *Pharmacological Basis of Therapeutics*. Tenth Edition. New York. Macmillan Company, pp. 687-720.

Ronald R.B. 2002. Non-steroidal anti-inflammatory agents, in Foye W.O. Ed. *Principles of Medicinal Chemistry,* 5th edition, Williams and Wilkins, N.Y.

Rotruck J.T., Pope A.L., Ganther H., Awanson A.B., Hafeman D.G. and Hoeckstra W.G. 1973. Selenium biochemical role as a component of glutathione peroxidase. *Science* 179: 588-590.

Sinha K.A. 1972 colorimetric assay of catalase. *Anal Biochem.* 47: 389-394.

Srinivasan K., Muruganathan S., Lal J., Chandra S., Tandan S.K. and Prakash V.R. 2001. Evaluation of anti-inflammatory activity of *Pongamia pinnata* leaves in rats. *J. Ethanopharmacol* 78: 151-157.

Steinmeyer J. 2000. Pharmacological basis for the therapy of pain and inflammation with nonsteroidal anti-inflammatory drugs. *Arthritis*. Res. 2: 379-385.

Usoh I.F., Akpan E.J., Etim E.O. and Farombi E.O. 2005 Antioxidant actions of dried flower extracts of *Hibiscus sabdarifa* L. on sodium arsenite induced oxidative stress in rats. *Pak J. Nutrition* 4 (3): 135-141.

Vane J.R., Bable Y.S. and Blotting R.M. 1998. Cyclooxygenase 1 and 2. *Annu. Rev. Pharmacol. Toxicol.* 38: 97-120.

Wang S.Y. and Jiao H. 2000. Scavenging capacity of berry crops on superoxide radicals, hydrogen peroxide, hydroxyl radicals, and singlet oxygen. *J Agri Food Chem* 48: 677-684.

Zielinski H. and Kozlowska H. 2000. Antioxidant activity and phenolics in selected cereal grains, their different morphological fraction. *J Agri Food Chem* 48: 2008-2016.

10

Biodiversity Conservation and Traditional Knowledge

Pratibha Sharma, Hemen Sarma and *C.M. Sarma*

1. Introduction

India has an amazing biodiversity, a result of its varied climate, soil and its rich cultural and ethnic diversity. At least 166 species of crops and 320 species of wild relatives of crops are known to have originated in India. An estimated 50,000 to 60,000 varieties of rice grow in India. Medicinal plants are being used as bio-cultural resources traditionally in our country since past many centuries. There are over 53 million tribal people belonging to 50 communities. Nearly 80 percent of the Northeastern States like Meghalaya, Nagaland, Mizoram and Arunachal Pradesh belong to tribal communities. The central Indian tribal belt of Madhya Pradesh, Orissa, Bihar, Gujarat and Rajasthan also has rich genetic diversity. India is one of the twelve-mega biodiversity of the world. Northeast India is amongst one of the 34 biodiversity hot-spots recognized in the world. This particular region is endowed with a rich wealth of medicinal and aromatic plants. This vital flora is now being regarded as a 'natural treasures' having a continuing social and growing economic importance. Since times immemorial plants have been utilized for human and veterinary health.

90% of herbal drugs used in the manufacture of Ayurveda, Siddha, Unani and Homeopathy systems of medicine are largely obtained from the wild. This wild resource is speedily shrinking day-by-day. Medicinal and Aromatic Plants (MAPs) are being utilized and traded in the lower Himalayan regions and plains of India from the Higher Himalayan mountains. The international trade in medicinal plants is estimated at US $ 70 billion and is growing at a rate of seven per cent annually. The Planning Commission of the Government of India has planned to increase trade in medicinal plant extracts to

Rs. 3,000 crores by the year 2005 and Rs. 10,000 crores by the year 2010. The importance and global commitment towards the conservation of biodiversity is no longer questioned. The accelerating rate of floral and faunal species loss and the projected negative impacts of this loss of germplasm on humankind have been eloquently described by a growing number of prominent biological scientists. Given their vital utility on one hand and threats to their survival on the other, it makes it imperative to plan for their conservation alongside promotion. There is a need to regulate the harvest of medicinal plants from the wild for commercial exploitation. Therefore, there is a need for conservation and sustainable use of medicinal plants. Traditional knowledge is vital for sustainability of natural resources including forests, water, and agro ecosystems across landscape continuum spanning from households through farms, villages and wilderness. Contemporary research on traditional and formal knowledge systems demonstrates the value of traditional knowledge for biodiversity conservation.

2. Indigenous Knowledge and Biodiversity

During the past decade progress has been made in understanding the complementarily of cultural diversity and biodiversity. The *ex-situ* conservation of germplasm has been well-established through the Plant Introduction Stations in the United States, the Consultative Group on International Agricultural Research and the International Board for Plant Genetic Resources. A parallel set of institutions documenting local knowledge about the environment is now being established. Indigenous knowledge resource centers could play a future role in *in situ* conservation. Understanding the indigenous management strategies of farmers and other rural persons that foster diversity in domesticated and wild species can help in the establishment of national programs for *in situ* conservation of germplasm that complement the *ex-situ* programs already in existence (Wilkes 1991; Altieri & Merrick 1988; Juma 1989).

National Research Council, focused on the conservation of biodiversity and on sustainability issues in agriculture and natural resource management. The document outlining USAID's newest Collaborative Research Support Program for Sustainable Agriculture and Natural Resource Management (SANREM) recognizes that indigenous knowledge must play an important role in sustainable approaches to development (National Research Council, 1991).

3. Traditional Knowledge on Biodiversity Conservation

Protection of knowledge, innovations and practices associated with biological resources do not seem to fall in the conventional legal systems of IPR protection (e.g. patents, copyright, trademark etc.). These conventional forms of IPRs are adequate to protect indigenous knowledge essentially because they are blessed on protection of individual property rights, thus traditional knowledge is, by and large, a collective term. Further, the informal knowledge exhibits other difficulties on being recognized for the purpose of IP protection, such as:

- Knowledge is developed over a period of time and may either be codified in texts or retained in oral traditions over generations. The conditions of novelty and innovative step necessary for grant of patent are therefore not satisfied.
- Communities quite often hold knowledge in parallel

Nevertheless, the development of an appropriate form of protection for the knowledge of local communities is of great interest to countries which are rich in biodiversity and also rich in traditional knowledge, such as India.

Documentation of Traditional Knowledge

It is believed that proper documentation of associated traditional knowledge could help in checking bio-piracy. It is assumed that if the material/knowledge is documented, it can be made available to patent examiners the world over so that prior art in the case of inventions based on such materials/knowledge are/is readily available to them. It is also hoped that such documentation would facilitate tracing of indigenous communities with whom benefits of commercialization of such materials/knowledge has to be shared. On the other hand, documentation may facilitate bio-piracy.

Traditional Knowledge Digital Library

In the recent past, there have been several cases of bio-piracy of traditional knowledge from India. For preventing such instances in future there is a need for developing digital databases of prior art related to herbs already in the public domain. In India, an exercise has been initiated to prepare easily navigable computerized database of documented traditional knowledge relating to use of medicinal and other plants (which already under public domain) known as traditional knowledge digital library. Such digital database would enable Patent Officers all over the world to search and examine any prevalent use/prior art, and thereby prevent granting of such patents and biopiracy.

Registration and Innovation Patent System

This involves creating a system for registration of innovations by inventors. Such registration will be tantamount to giving right to the inventor to challenge any use of the innovation without prior permission. For novel and useful innovations, some kind of petty patent giving protection for a limited duration may be worked out.

Development of a *Sui-generis* System

Some experts have suggested that a *sui-generis* system separate from the existing

IPR system should be designed to protect knowledge, innovations and practices associated with biological resources. However the parameters, elements and modalities of a *sui-generis* system are still being worked out. This is in addition to the *sui-generis* system of protection for plant varieties. The *sui-generis* system of protection for plant varieties is separately developed in India and a bill in this regard is to be tabled in the Parliament.

Local Vegetation Management: Local people have developed a variety of vegetation management practices over thousands of years that continue to exist in tropical Asia (Pandey, 1998), South America (Atran *et al.* 1999; Gomez-Pompa and Kaus, 1999), Africa (Getz *et al.* 1999; Infield, 2001) and other parts of the world (Brosius, 1997; Berkes, 1999). People also follow ethics that often help them regulate interactions with their natural environment (Callicott, 2001).

In India these systems can be classified in several ways:

- Religious traditions: temple forests, monastery forests, sanctified and deified trees
- Traditional tribal traditions: sacred forests, sacred groves and sacred trees
- Royal traditions: royal hunting preserves, elephant forests, royal gardens etc.
- Livelihood traditions: forests and groves serving as cultural and social space and source of livelihood products and services

The traditions are also reflected in a variety of practices regarding the use and management of trees, forests and water. These include:

- Collection and management of wood and non-wood forest products
- Traditional ethics, norms and practices for restraint use of forests, water and other natural resources
- Traditional practices on protection, production and regeneration of forests.
- Cultivation of useful trees in cultural landscapes and agro forestry systems
- Creation and maintenance of traditional water harvesting systems such as tanks along with plantation of the tree groves in the proximity

These systems support biodiversity, which is although less than natural ecosystems but it helps reduce the harvest pressure. In India, local practices of vegetation management perhaps emanate from the basic ecological concepts of local communities reflected in "ecosystem-like concepts in traditional societies" (Berkes *et al.* 1998).

4. Biodiversity in Sacred Cliffs

Cliffs are completely forgotten cultural landscape elements that support a variety of species of plants and animals in India. As humans have special fascinations to such areas often cliffs across the country are considered sacred. Cliffs elsewhere have been found to support undisturbed ancient woodland, dominated by tiny, slow-growing and widely spaced trees. Vertical cliffs often support populations of widely spaced trees that are exceptionally old, deformed and slow growing. Some of the most ancient and least-disturbed wooded habitats on Earth are found on cliffs, even if such sites are close to intensive agricultural and industrial development.

5. Farm Biodiversity

In the Indian farms and fields one finds strips of vegetation containing several species of plants and small animals. These strips are beneficial in several ways. Such strips on tropical lands have been found to accelerate natural successional processes by attracting seed-dispersing animals and increasing the seed rain of forest plants. Effects of these strips resemble the windbreaks on seed deposition patterns (Harvey, 2000). Isolated trees provide seed in the area for natural regeneration. The strips enhance seed rain and connectivity. Because such strips trap large number of seeds of several species they help in further tree growth. Compared to open fields, farm boundaries with vegetation receive seeds in greater densities and species-richness than open farms and pastures. All forms of seed dispersal help in the process but animal-dispersed (birds, bats, mammals etc.) seeds often occur in greater densities and species number. Farm boundaries maintained throughout the country are often self-regenerating and require only management as these barriers considerably increase the deposition of tree and shrub seeds within the cultural landscape. Indeed considerable biodiversity is found within these strips. This is a practice that needs to be maintained as it has several socio-economic benefits as well.

6. Cultivation of Medicinal Plants

In India there are numerous examples of medicinal plant cultivation by local people. Socio-culturally valued species find place in home gardens and courtyards. Interestingly, the annual production of medicinal plants has been found to be comparable with the annual production of traditional crops. Thus, cultivation and harvesting can help in livelihood security and *in situ* conservation of these species.

This suggests a wealth of traditional knowledge on biodiversity and herbal health care in tribes of eastern India. Tribes in the region are dependent on forests for other species as species of mushrooms, wild berries, tubers, and flowers that are included in their diet including cooking oil. Understanding of traditional knowledge on biodiversity of the region will be very helpful in planning for sustainable forest management.

Traditional Ethos

In spite of the modernization, traditional ecological ethos continues to survive in many local societies, although often in reduced forms. Investigations into the traditional resource use norms and associated cultural institutions prevailing in rural Bengal societies (Deb & Malhotra, 2001) demonstrate that a large number of elements of local biodiversity, regardless of their use value, are protected by the local cultural practices. Some of these may not have known conservation effect, yet may symbolically reflect a collective appreciation of the intrinsic or existence value of life forms and the love and respect for nature. Traditional conservation ethics are still capable of protecting much of the country's decimating biodiversity, as long as the local communities have even at stake in the management of natural resources.

One example from Northeast India is particularly notable (Tiwari *et al.* 1998). The tribal communities of Meghalaya – Khasis, Garos and Jaintias have a tradition of environmental conservation based on various religious beliefs. As elsewhere in India, particular patches of forests are designated as sacred groves under customary law and are protected from any product extraction by the community. Such forests are very rich in biological diversity and harbor many endangered plant species including rare herbs and medicinal plants.

Rural Women as Protectors

In many ways, the relationship between women and forest resources parallels that of rural women and agriculture. Trees and forest ecosystems are to the rural women of developing countries as what the supermarket or hardware store is for those in industrialized countries, says a forest official. For many landless women, the only source of income is often just the forest produce. Women are dependent on forests for food, fuel, fodder, medicines, fibers and so on. Though women can play a significant role in protecting biodiversity, it has not been recognized. Their role in conservation has been over looked. The loss of habitats and biodiversity mostly affects the underprivileged and many of them are women.

7. Integration of Traditional and Formal Science

Traditional knowledge may indeed complement scientific knowledge by providing practical experience in living within ecosystems and responding to ecosystem change. Indian traditions and local knowledge have often paved the way for many discoveries in science. Traditional skills, local techniques and rural craft provide a wide spectrum of knowledge in India. We have to take the validated local knowledge into account together with science for evolving a robust sustainability science. Sharp boundaries between formal and local systems of knowledge, natural sciences and social sciences may indeed be imaginary.

8. Incorporating Traditional Knowledge in Practice

Any attempt, endeavouring to integrate traditional knowledge for biodiversity conservation and sustainability of natural resources should be based on the principle that traditional knowledge often cannot be dissociated from its cultural and institutional setting. Regarding the cultural and institutional setting, the following suggestions may be useful:

1. Each programme aiming at the promotion of traditional knowledge should be based on the recognition that natural resource rights and tenurial security of local communities forms the fundamental basis of respecting traditional knowledge.
2. More attention is needed on protection of intellectual property rights of traditional people.
3. Innovative projects may need to be developed that aim at the enhancement of the capacity of local communities to use, express and develop their traditional knowledge on the basis of their own cultural and institutional norms.

There is an urgent need for the integration of traditional and formal sciences. Following considerations may be useful in this regard:

1. Development of methods for mutual learning between local people and the formal scientists.
2. State forest policies and sustainable forest management processes need to give full attention to the forestry and local institutional arrangements to incorporate traditional knowledge in forest management and development projects.
3. Traditional knowledge and traditions can contribute to the preparation of village microplans, which are prepared for eco-development, joint forest management and rural development. The plans should be based on both geographic and traditional community boundaries rather than only on administrative boundaries.
4. Revival of the traditional water management systems that have served the society for hundreds of years but are currently threatened
5. There is a clear need to integrate traditional and formal sciences for participatory monitoring and taking feedback to achieve adaptive strategies for management of natural resources.

In spite of the value of traditional knowledge for biodiversity conservation and natural resource management, still there is a need to further the cause. The following considerations may be useful in this respect:

1. Encouraging the documentation of indigenous knowledge and its use in natural resource management. Such documentation should be carried out in participation with the communities that hold the knowledge. Due attention should be given to document the endemic perspectives regarding IK rather than only the perspectives of professional outsiders
2. Facilitating the translation of available and new documents describing Indic traditions such as ancient texts on medicinal plants into local languages and dissemination of these documents amongst local people. Such a translation is indeed required because texts are often available in languages (e.g. Sanskrit) not understood by many in contemporary India. On the other hand, translation of local knowledge into formal scientific terminology will provide space to external researchers, policy makers and practitioners to comprehend and support people's knowledge systems and initiatives.
3. Facilitating the exchange of information amongst practitioners of local knowledge.
4. Developing clear and concise educational material on traditional knowledge systems to be used in communication programmes to impart information regarding the merits and threats to indigenous knowledge systems to both policy makers and the general public.

Scientific institutions have an important role to play in supporting the knowledge systems. As has been pointed out earlier, it is now recognized that a dichotomy between local and formal systems of knowledge is not real and that any knowledge is based on a set of basic values, beliefs and paradigms. Therefore, there is a definite need to further develop systematic insight into the nature and scope of traditional knowledge. The following activities may be useful in this regard:

1. Developing curricula and methods for providing formal training and education in traditional knowledge systems to agencies, researchers and practitioners who work in collaboration with communities.
2. Developing research projects aimed at assessing the possibilities and constraints of using traditional knowledge under specific conditions. Such research projects should move beyond the first generation research projects, which aimed at demonstrating the value of local knowledge systems by focusing on successful cases of application. Second generation research projects shall focus on comparing

application of knowledge systems across a range of circumstances and across disciplines to craft the traditional sustainability science.

3. Developing new methods for incorporating local knowledge systems in natural resource management regimes through action research.

REFERENCES

Altieri M. A. and Laura C. Merrick. 1988. Agroecology and In Situ Conservation of Native Crop Diversity in the Third World. *In Biodiversity*, E. O. Wilson and F. M. Peter, eds. Pp. 15-23. Washington, D.C.: National Academy Press.

Atran S. *et al.* 1999. Folkecology and commons management in the Maya Lowlands. *Proceedings of the National Academy of Sciences USA* 96: 7598-7603.

Berkes F., Kislalioglu M., Folke C. and Gadgil M. 1998. Exploring the basic ecological unit: Ecosystem-like concepts in traditional societies. *Ecosystems* 1: 409-415.

Berkes F. 1999. *Sacred Ecology: Traditional Ecological Knowledge and Resource Management.* Taylor and Francis, Philadelphia.

Brosius P.J. 1997. Endangered forest, endangered people: environmentalist representations of indigenous knowledge. *Human Ecology* 25: 47-69.

Callicott J. B. 2001. Multicultural environmental ethics. *Daedalus* 130(4): 77-98.

Deb D. and Malhotra K. C. 2001. Conservation ethos in local traditions: the West Bengal heritage. *Society and Natural Resources* 14: 711-724.

Getz W. M., Fortmann L., Cumming D., Toit J. du., Hilty J., Martin R., Murphree M., Owen-Smith N., Starfield A.M. and Westphal M.I. 1999. Sustaining natural and human capital: villagers and scientists. *Science* 283: 1855-1856.

Gomez-Pompa A. and Kaus A. 1999. From pre-Hispanic to future conservation alternatives: lessons from Mexico. *Proceedings of the National Academy of Sciences USA* 96: 5982-5986.

Harvey C. A. 2000. Windbreaks enhance seed dispersal into agricultural landscapes in Monteverde, Costa Rica. *Ecological Applications* 10: 155-173

Infield M. 2001. Cultural values: a forgotten strategy for building community support for protected areas in Africa. *Conservation Biology* 15: 800-801.

Juma Calestous. 1989. *Biological Diversity and Innovation: Conserving and Utilizing Genetic Resources in Kenya.* Nairobi: African Centre for Technology Studies.

Kunte K., Joglekar A., Utkarsh G. and Padmanabhan P. 1999. Patterns of butterfly, bird and tree diversity in the Western Ghats. *Current Science* 77: 577-586.

Larson D. W., Matthes U. and Kelly P. E. 2000b. *Cliff Ecology.* Cambridge University Press, U.K.

National Research Council. 1991. *Toward Sustainability: A Plan for Collaborative Research on Agriculture and Natural Resource Management*. Washington, D.C.: National Academy Press.

Pandey D.N. 1998. *Ethnoforestry: Local Knowledge for Sustainable Forestry and Livelihood Security*. Himanshu/AFN, New Delhi.

Peterken G. F. 1996. *Natural Woodland*. Cambridge University Press, UK.

Rawal R.S. and Dhar U. 2001. Protected area network in Indian Himalayan region: Need for recognizing values of low profile protected areas. *Current Science* 81: 175-184.

Tiwari B. K., Barik S. K. and Tripathi R. S. 1998. Biodiversity value, status, and strategies for conservation of sacred groves of Meghalaya, India. *Ecosystem Health* 4: 20-32.

Wilkes Garrison. 1991. *In Situ* Conservation of Agricultural Systems. In *Biodiversity: Culture. Conservations, and Ecodevelopment*, Margery L. Oldfield and Janis B. Alcorn,eds. Pp. 86-101. Boulder: Westview Press

Wilson., E. O. 1998. *Consilience: The Unity of Knowledge*. Knopf, New York.

11

Study on the Adaptogenic Property of *Clerodendron colebrookianum* Walp. on Cold Stress Induced-Mice

R.M. Syiem, A. Chatterjee, S. Majaw and *S. R. Joshi*

1. Introduction

Stress is a state of altered homeostasis, resulting from an external or internal challenge *i.e.* a stressor and is characterized by activation of both the hypothalamic-pituitary-adrenal (HPA) axis and the sympathetic nervous system (SNS) (Kizaki *et al.* 1997). Environmental stimuli like pain, hunger, thirst or cold disrupts homeostasis and results in physiological changes in the animal and is related to the development of a pathological state (Selye, 1932). Stress has been postulated to be involved in the etiopathogenesis of a variety of disease states, including hypertension, peptic ulcer, diabetes, immuno-suppression, reproductive dysfunctions (Staratakis *et al.* 1995) and behavioral disorders like anxiety (Hata *et al.* 2000) due to involvement of the central nervous system (CNS), endocrine system and metabolic system. The resultant disturbances may vary due to the type, intensity and duration of a particular stressor and the strain/sex differentiation of the experimental subjects (Kioukia-Fougia *et al.* 2002).

Exposure to cold is a direct threat to the body. It was observed that the specific changes in body temperature of ± 4°C from normal could impair both physical and mental task (Astrand *et al.* 1970). When the surrounding temperature drops below 18°C, the body may not be able to warm itself and hence serious cold-related illnesses, permanent tissue damage and death may result (NIOSH, 2002).

Adaptogens are plant extracts that allow an organism to counteract adverse physical, chemical and biological stressors by generating nonspecific resistance. Adaptogens are known to increase the availability of energy during the day, reduce stressed feelings, increase endurance and increase mental alertness (Farhath *et al.* 2005).

Certain plants like *Boerhaavia diffusa, Ocimum sanctum, Centella asiatica, Withania somnifera, Eleuthrococcus senticocus, Panax ginseng, Raponticum carthamoides* and *Rhodiola rosea,* which are used as tonics in folk medicine, have been demonstrated to fulfill the criteria of plant adaptogens (Singh *et al.* 2001, Pandey *et al.* 2005). Most adaptogenic plant species are reported to have estrogenic, tonic, antioxidant, nootropic and aphrodisiac properties. These properties can be linked, however, to the neuroendocrine systems involved in response to stress (Klein, 2006). The present study was aimed to investigate the anti-stress activity of the aqueous extract of *Clerodendron colebrookianum* Walp. leaves on mice. This was done by measuring WBC count and change in spleen weight.

2. Materials and Methods

Test animals

Adult male Swiss albino mice weighing between 20-30g were used for the study. Mice were housed in a room kept under controlled conditions with temperature maintained at 22°C on a 12 h light/dark cycle and were fed with balanced mice feed (Pranav Agro Industries Ltd India).

Plant material

Clerodendron colebrookianum Walp. leaves were collected from different locations of Meghalaya.

Extraction

C. colebrookianum leaves were washed, shredded and dried. The leaves (250g) were then powdered, homogenized and repeatedly extracted with 10 volumes (2.5L) of water (Syiem *et al.* 2002), the mixture was filtered and the filtrate was lyophilized. The dried mass obtained was used for the investigation. Prior to use, weighed powder was dissolved in water and the mixture was used for further study.

Induction of stress

The mice were individually placed in a well-aerated jar (height = 16cm, diameter = 6cm). The containers were placed in a cold chamber maintained at 4°C for 3 hours per day for 5 consecutive days (Sumanth *et al.* 2005).

Administration of extract

The lyophilized extract was weighed and dissolved in 100ml of distilled water (100mg/Kg body weight). This extract preparation was then given to mice orally by substituting the extract in place of water. The mice were allowed to take this extract for a period of 7 days prior to stress and 5 days during the stress treatment.

Collection of Blood and Blood analysis

Blood samples from the control and experimental mice were collected by orbital sinus puncture using heparinized capillary glass tubes (Ivorra *et al.* 1988). The blood samples so collected were analyzed for WBC count using Neubauer's chamber (Marienfeld, Germany).

Weighing of organs

The mouse was dissected in the abdomen; the spleen was taken out and weighed (Brekhman, 1969; Kim *et al.* 1979).

3. Results

White Blood Cells (WBC) Count

It was observed that there was a reduction in the WBC count of stress mice compared to the control mice on completion of 5 days of stress treatment. On administration of the extract it was observed that the WBC count was almost to the level of the control mice (Fig.1).

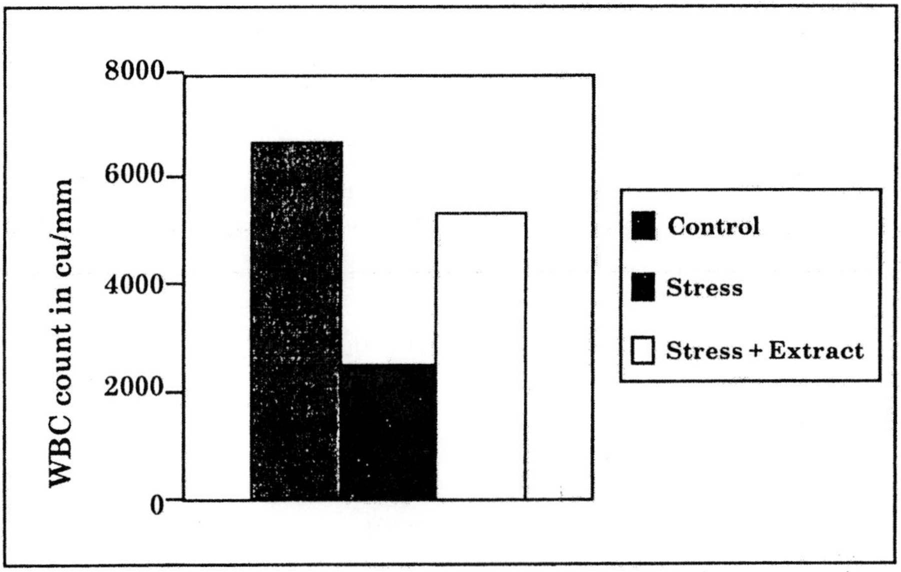

Figure1. Effect of extract on WBC count of stress induced mice. Values taken as mean (Number of samples 3-6 mice).

Weight of Internal Organs

There was a reduction in spleen weight after the cold-stress treatment. When extract was given to mice it was found that, there was no significant decrease in the spleen weight and was almost close to that of the control mice (Fig.2)

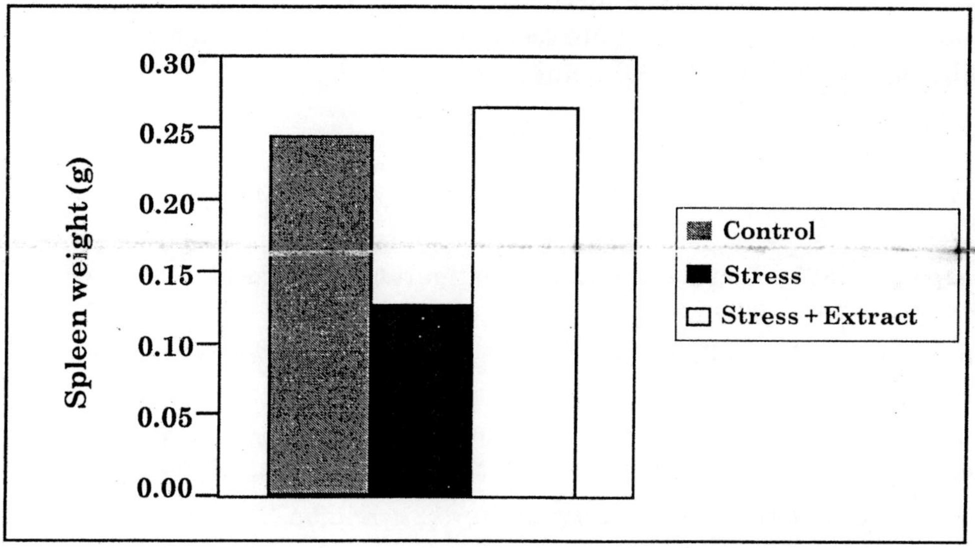

Figure 2. Effect of extract on spleen weight after 5 days of stress treatment. Values taken as mean (Number of sample 3-6 mice)

4. Discussion

From the study, it is evident that cold stress can cause a number of physiological damages to the organism. Cold stress leads to the release of steroid hormone 'Glucocorticoid' from the adrenal cortex which has different effects on carbohydrate metabolism, immune system, protein metabolism and fat metabolism (Guyton *et al.* 2001; Ganong, 2003).

It was found that the cold stress treatment reduces the White Blood Cell (WBC) count and this reduction could be due to the lympholysis activity of the Glucocorticoid on immature lymphocytes and it also leads to sequestration of lymphocytes on the lymphoid organs and as a result the WBC count drops down in the circulation (Ganong, 2003; Pineda *et al.* 2003; Coico *et al.* 2003). Cold stress also leads to shrinkage or decrease in the weight of spleen. The weight of lymphoid organs like spleen usually decreases with abnormal Glucocorticoid level because high level of this hormone leads to widespread lympholysis in the lymphoid organs (Goldsby *et al.* 2003).

The extract of *C. colebrookianum* was fed to the mice to check if there is any antistress activity present in the extract. It was found that the extract was able to maintain WBC count to the level of the control mice. The extract was also able to maintain the normal weight of the internal organs i.e., spleen. With these insufficient data on the given extract, it will be difficult to predict the mechanism of action and the active component(s) involved which are responsible for the antistress activities. But these results would provide basic data to examine antistress effects of this plant and further indepth studies could contribute in the development of functional plant materials with antistress activities through characterization of active plant metabolite(s).

REFERENCES

Astrand P. O. and Rodahl K. 1970. *Textbook of Work Physiology*, McGraw-Hill Book Company, New York, U.S.A.

Brekhman II. 1969. Pharmacological investigation of glycosides from Ginseng and *Eleutherococcus*. Lloudia 32:16.

Coico R., Sunshine G. and Benjamini E. 2003. *Immunology A Short Course* (Fifth Edition), Wiley – Liss, USA.

Farhath K., Amarinder S.B. and Brahm S. 2005. *Rhodiola rosea*: A Versatile Adaptogen. *Comprehensive Reviews in Food Science and Food Safety* 4: 55-62.

Ganong W.F. 2003. *Review of Medical Physiology*, (21st Edition), McGraw Hill.

Goldsby R.A., Kind T.J., Osborne B.A. and Kuby J. 2003. *Immunology*. (Fifth edition), W.H Freeman and Company, New York.

Guyton A.C. and Hall J.E. 2001. *Text of Medical Physiology*, (Tenth Edition), Saunders Hartcourt India.

Hata T., Itoh E. and Nishikawa H. 2000. Stress-induced anxiety and endogenous anxiogenic substances. *Japanese Journal of Pharmacology*, 115: 13– 20.

Ivorra M.D., Paya M. and Villar A. 1988. Hypoglycemic and insulin release effects of tormentic acid, a new hypoglycemic natural products. *Planta Med* 54: 282-286.

Kim N.D., Hamn B.H., Lee E.B., Kong J.Y., Kim M.H. and Jin C.B. 1979. Studies of ginseng on the antistress effects. *Korean J Pharmacog* 10: 61-67.

Kioukia-Fougia N., Antoniou K., Bekris S., Liapi C., Christofidis I. and Papadopoulou-Diafoti Z. 2002. The effects of stress exposure on hypothalamic-pituitary-adrenal axis, thymus, thyroid hormones and glucose levels. *Progress in Neuropharmacology and Biological Psychiatry*, 26: 823–830.

Kizaki T., Ookawara T., Izawa T., Nagasawa J., Haga S., Radák Z. and Ohno H. 1997. Relationship between cold tolerance and generation of suppressor macrophages during acute cold stress. *Journal of Applied Physiology*, 83: 1116-1122.

NIOSH (National Institute for Occupational Safety and Health). 2002. *Safety notes*, 2: pp. 22.

Pandey R., Maurya R., Singh G., Santhiamoorthy B. and Naik S. 2005. Immunosuppressive properties of flavonoids isolated from *Boerhaavia diffusa* Linn. *International Immunopharmacology*, 5: 541-553.

Pineda M.H. and Dooley M.P. 2003. *Veterinary Endocrinology and Reproduction*, (Fifth Edition), Iowa State Press, Ames Iowa.

Robyn K. 2006. Phylogenetic and phytochemical characteristics of plant species with adaptogenic properties MS Thesis, 2004, Montana State University. www.rrreading.com/files/Thesis%20Table%20Abstract.pdf

Selye H. 1932. The general adaptation syndrome and the diseases of adaptation. *Journal of Clinical Endocrinology*, 6: 117-152.

Singh B., Saxena A.K., Chandan B.K., Gupta D.K., Bhutani K.K. and Anand K.K. 2001. Adaptogenic activity of a novel, withanolide-free aqueous fraction from the roots of *Withania somnifera* Dun. *Phytotherapy Research*, 15: 311-318.

Staratakis C.A. and Chrousos G.P. 1995. Neuroendocrinology and pathophysiology of the stress system. *Annals of the New York Academy of Sciences*, 771: 1–18.

Sumanth M. and Mustafa S.S 2005. Antistress, Adoptogenic activity of *Aegle marmelo* fruit in Mice Review.

Syiem D., Gareth Syngai., Khup P.Z., Khongwir B.S., Kharbuli B. and Kayang H. 2002. Hypoglycemic effect of *Potentilla fulgens* L. in normal and alloxan-induced diabetic mice. *J Ethnopharmacol* 83: 55-56.

12

Ex-Situ Conservation and Sustainable Utilization of Medicinal Plants at Dibrugarh, Assam

Sristisri Upadhyaya and L.R. Saikia

1. Introduction

Medicinal plants have been crucial in sustaining the health and well-being of mankind. They are the local heritage with global importance. The world is endowed with a rich wealth of medicinal plants. It is estimated that around 70,000 plant species have been used at one time or another for medicinal purposes. In India, medicinal plants have made a good contribution to the development of ancient Indian Materia Medica. Moreover, people are increasingly gaining faith in Indian system of medicines like Ayurveda, Unani and Siddha. About 80% of the populations of many developing countries still use traditional medicines for their health care. The demands of the medicinal plants have been met by indiscriminate harvesting of spontaneous flora. As a result many plant species have become extinct and some are endangered. It is therefore necessary that systematic cultivation of medicinal plants be introduced in order to conserve biodiversity and protect threatened species.

As far as the procurement, collection, cultivation, sale, purchase, import and export of the medicinal herbs are concerned, as such, there is no definite procedure and no scientific data are available in the country. Judicious collection, multiplication, preservation and utilization are perhaps the best policy for conservation of medicinal plants. To achieve the goal we should not go for conventional method of management for this group of plants.

The present communication deals with (i) collection of medicinal plants having immense potentialities, their plantation and preservation (germplasm conservation) in the Botanical garden of Life Sciences Department of Dibrugarh University, (ii) scientific study of some other wild species about their utilization as folklore medicine by the people around the Jakai Reserve Forest, and (iii) scientific cultivation of five potential medicinal plants species viz. *Acorus calamus, Andrographis paniculata, Asparagus racemosus, Adhatoda vasica* and *Centella asiatica.*

2. Materials and Methods

The study has been conducted at Dibrugarh (27° N and 94° E) above 150 m mean sea level. The soil of the area is alluvial and acidic in nature (p^H 5.3 to 6.5). The climate is tropical monsoon type with broad seasons. The summer season extends from mid March to mid October and is characterized by high rainfall, high humidity and high temperature (max. aver. 33°c). The winter season (late Oct. to early March) is practically rainless and average temperature goes down to 12°C.

Forty eight medicinal herbs native to the garden are grown well in protected condition. Another forty three potential medicinal species were collected by field visit to nearby native and reserved forests (Jokai Reserve Forest, Joypur Reserve Forest) through the officials of Jokai Botanical Garden and also by personal contact with the knowledgeable persons of the area.

To know the uses of these plants as folklore medicine, we have visited the inhabitants in and around the reserve forest and information was gathered by personal contact with the elderly persons and practitioners.

A scientific cultivation practice (including nursery development) of five potential species was started in the month of August 2006 and the plants were transplanted in the experimental plots during March 2007. Cow dung, FYM, compost and vermicompost were used as manure and watering was done regularly during sunny days.

Biopesticides prepared from garlic and tulsi have been used to control Looper caterpillar. 20 gram crushed garlic was thoroughly mixed with 1 liter boiled water and 0.5 gram black piper powder are added to it. A nut size soap was added and shaked well before spraying. Another biopesticide prepared was tulsi leaf leachate (10% wet wt) and soap was added as emulsifier. Both biopesticides were applied in separate plots in alternate days.

3. Results

The photographs of certain potential species are taken in the last part of July 2007 and have been prepared for power point presentation along with a brief description of each species. The species are *Acorus calamus, Adhatoda vasica, Alternanthera sessilis, Aloe vera, Amorphophalus campanulatus, Andrographis paniculata, Anthocephalus cadamba, Aquilaria agallocha, Argemone mexicana, Asparagus racemosus, Averrhoa carambola, Azadirachta indica, Bauhinia recemosa, Bombax ceiba, Butea monosperma, Bacopa monnieri, Bryophyllum pinnatum, Baccaurea sapida, Capsicum pendulatum, , Catheranthus roseus, Centella asiatica, Cassia fistula, Cissus quadrangularis, Curcuma anguistifolia, Clerodendron colebrookianum, Coleus forskohlii, Costus speciosus, Drymeria cordata, Emblica officinalis, Eugenia jambolana, Euphorbia hirta, E. neriifolia, Garcinia pedunculata, Hydrocotyl rotundifolia, Houttuynia cordata, Jatropha curcus, Ocimum sanctum, O. basilicum, Lasia spinosa, Leucas aspera, Mallotus albus, Mesua ferrea, Moringa oleifera, Morus indica, Myrica nagi, Murrya koeningii, Mussaenda frondosa, Nyctanthes arbor-tristis, Nymphaea esculanta, Oldanlandia corymbosa, Oroxylum indicum, Ocimum sanctum, O. basilicum, Paederia foetida, Perilla frutescence, Portulaca oleraceae, Phyllanthus niruri, Phlogocanthus thyrsiflorus, Piper longum, Pongamia pinnata, Pogostemon cablin, Psidium guyava, Polygonum microcephallum, Rouvolfia serpentina, Smilax macrophylla, Stevia rebudiana, Solanum torvum, S. nigrum, Scoparia dulcis, Terminalia arjuna, T. chebula, Tinospora cordifolia, Vitex nigundo, Wedelia calendulacea, Withania somnifera, Zingiber perpureum.*

Eighty nine species used as folklore medicines are presented in Table 1 along with their uses and the part(s) used in traditional healthcare system. Generally the medicines are given in three forms – fresh juice, decoction and prepared (cooked). Extracts and leachates are prepared occasionally. Some of the important method of preparation of drugs would be mentioned during oral presentation. Photographs of scientific cultivation of species are also presented in power point along with their brief description.

Table 1
Medicinal plants used as folklore medicine by the local people

Species	Part(s) used	Ailments
Acorus calamus Linn.	rhizome	Bronchial cattarrah
Adhatoda vasica Nees.	leaf, root	Cough, bronchitis, asthma, dysentery
Aegle marmelos Correa	fruit	Dysentery, Diarrhoea
Aloe vera Linn	leaf	Fever
Alstonia scholaris (L)R. Br.	bark	Diarrhoea, Dysentery, Malaria, Leprosy
Alternanthera sessilis R. Br.	whole plant	Dysentery, Diarrhoea

(Contd...)

Species	Part(s) used	Ailments
Amorphophalus campanulatus (Roxb) Blume	Corm	Piles, dysentery
Andrographis paniculata Nees.	whole plant	Fever, cough, stomach trouble
Anthocephalus cadamba Roxb.)Miq.	bark	Fever, cough
Argemone mexicana Linn.	latex	Jaundice
Asparagus racemosus Willd.	tuberous root	Urinary trouble, rheumatism,
Aquilaria agallocha Roxb.	bark	Diarrhoea, stop vomiting, rheumatism.
Averrhoa carambola Linn.	fruit	Jaundice
Azadirachta indica Juss.	leaves	Small pox, insect repellant, worm repellant, skin disease
Bauhinia racemosa Lam.	leaf, bark	Headache, malaria, jaundice
Bombax ceiba Linn.	gum, bark	Diarrhoea, dysentery
Butea monosperma L.(Kuntz)	seed	Expulsion of round worms
Bacopa monnieri (L.) Pennell	whole plant	Brain tonic
Bryophyllum pinnatum Kuntz.	leave	kidney stone expeller, urinary trouble
Baccaurea sapida Muell and Arg.	fruit	Body coolant
Capsicum pendulatum Linn.	fruit	Gastric, dysentery
Catharanthus roseus(L)G.Don.	flower	Diabetes mellitus
Calotropis gigantea *R.Br.*	flowers, root, bark	Cough, asthma, catarrhact dysentery, cough
Coleus forskohlii (Willd)Brig	leaf	Digestive stimulant
Centella asiatica (L)Urbn.	whole plant	Diarrhea, dysentery
Cassia fistula Linn.	root	Fever, Heart disease
Cissus quadrangularis Linn.	stem	Bone injury
Curcuma anguistifolia Roxb.	rhizome	Digestive stimulator
Clerodendrum colebrookianum Walp	young shoot	High blood pressure
Costus speciosus (Koe) Smith.	rhizome	Jaundice
Drymeria cordata Willd.	leaf	Sinusitis
Emblica officinalis Geartn.	fruit	Jaundice, cough, eye and stomach trouble
Eugenia jambolana Lam.	bark, ripe fruit	Diarrhoea, dysentery, jaundice, diabetes

(Contd...)

Species	Part(s) used	Ailments
Euphorbia hirta Linn.	whole plant	Asthma, bronchitis
E. neriifolia Linn.	root, leaf	Antitode to snake poison, poison fish in river, cough
Garcinia morella Desv.	fruit	Dysentery, fever, jaundice, diabetes, urinary trouble
G. pedunculata Roxb.	fruit	Stomach trouble, digestive, dysentery
Houttuynia cordata Thunb.	whole plant	Indigestion, dysentery
Hydrocotyl rotundifolia Roxb.	whole plant	Diarrhoea, dysentery, brain tonic
Jatropha curcus Linn.	root bark	Rheumatism, gum, tooth ache
Lasia spinosa (L) Thw.	young leaf, root	White discharge of women, piles, sore throat
Leucas aspera Spreng.	twig, flower	Sinusitis, cold cough
Mallotus albus Muell.Arg.	bark	Cough
Mentha arvensis Linn.	young shoot	Jaundice, dysentery, to stop vomiting
Mesua ferrea Linn.	flower	Piles, burning sensation of feet
Mimosa pudica Linn.	root	Gum ache
*Mimosops elengi*Linn.	root bark	Gum and toothache fever
Moringa oleifera Lamk.	leaf, fruit, bark	Small pox, toothache, rheumatism
Morus indica Linn.	bark, root	Jaundice
Myrica nagi Thunb.	bark	Cough, tooth ache
Murrya koeningii Spreng.	leaf	Antioxidant
Mussaenda frondosa Linn.	root, bract	Leprosy, jaundice.
Nyctanthes arbor-tristis Linn.	leaf, flower	Rheumatism, expulsion of round and thread worms
Nymphaea esculanta Roxb.	corm	Expulsion of roundworm.
Oldanlandia corymbosa Linn.	whole plant	Fever, burning sensation, indigestion
O. diffusa Roxb.	root	Cold, cutaneous disorder, indigestion
Oroxylum indicum Vent.	bark	Anti carcinogenic
Ocimum sanctum Linn.	leaf, root	Cough, malarial fever
O. basilicum Linn.	leaves	Chronic diarrhoea, gonorrhea

(Contd...)

Species	Part(s) used	Ailments
Paederia foetida Linn.	whole plant	Abdominal pain, diarrhoea.
Piper longum Linn.	fruit, leaf	Chronic cough, child birth
Pogostemon cablin Benth	leaf	as insecticide
Perilla frutescence(L) Britt	twig	Abdominal pain, indigestion.
Portulaca oleraceae Linn.	whole plant	Urinary trouble
Phyllanthus niruri Linn.	whole plant	Jaundice
Plantago major L.	seed	Laxative, in constipation
Phlogocanthus thyrsiflorus Nees.	leaf, flower	Cough, worm trouble
Psidium guyava Linn.	twig, fruit	Diarrhoea, dysentery
Polygonum microcephallum Linn.	leaf	Stomach trouble
P.strigosum Br.	whole plant	Diarrhoea in cattle
Pongamia pinnata (L) Merr.	bark	Piles
Rauwolfia serpentina (L) Benth	root	Hypertension, labour pain
Smilex macrophylla Roxb	root	Rheumatism
Solanum indicum Linn.	root, leaf	Asthma, cough, toothache, vomiting
S. khasianum Clarke	seeds	Contraceptive
S.torvum Sw.	fruit	Enlarged spleen
S. nigrum Linn.	root, fruit	Asthma, hooping cough, piles
Scoparia dulcis L.	whole plant	Dry cough, sore throat
Sida rhombifolia Linn.	leaf, root	Tuberculosis
Tamarindus indica Linn.	leaf	Piles
Terminalia arjuna Wight & Arn.	bark	Heart disease
T. chebula Retz.	fruit, seed	Cough, asthma, piles
Tinospora cordifolia (Willd) Miers	Leaf, stem, root	Burn, diarrhoea, dysentery
Urena lobata Linn.	leaf, root	Tuberculosis
Wedelia calendulaceae Less.	leaf	Skin disease, hair tonic.
Withania somnifera Dunal	leaf	Fever
Vitex negundo Linn.	twig	Sinusitis, dullness of hearing, rheumatism
Zingiber perpureum Linn.	rhizome	Dry cough, indigestion
Zanthoxylum hamiltonium	bark, root	Rheumatism, toothache

(Contd...)

4. Discussion

The strategy adopted for conservation of MP has experienced that previously the area was very rich in MP biodiversity. Many of the local flora have been used by the local people from time immemorial. The remnants of many of the important MPs are there somewhere in the area but hitherto the scientific conservation strategy has not been adopted. All the 89 potential species reported here are frequently needed by the common people of the area in their traditional healthcare system but these species are not easily available (Saikia & Upadhyaya, 2006). Perhaps due to intensification of pressure on these species, they become vulnerable in the area. These bioresources have received attention of vendors, scientists, conservators and academics of present days (Sharma, 2004; Purohit & Vyas, 2005; Prajapati *et al.* 2006). The present approach seems to be an effective method of *in situ* conservation of these species since the germplasm of the species could establish and are growing well in the garden; thereby it becomes easy to conserve them in natural habitats.

These species need attention of conservationists due to over increasing pressure on them. People generally use these drugs in curing influenza, fever, cough, diarrhea, dysentery, stomach trouble, rheumatism, asthmatic trouble, bronchitis, diabetes, jaundice, malaria, toothache, bone injury etc. Perhaps this is how a large number of plant species have been discovered for their medicinal values. All these species are known to be very effective against the ailments according to the users and the practitioners. According to Jain (1994) the efficacy of the herbs depends on the total effect of the plant contents rather than on the one or two chemical fractions prepared from the herb. Although *A. agallocha* has primarily aromatic importance, the species is equally important as medicinal plant. Species like *A. calamus, O. sanctum, P. cablin* are also important as medicine and aromatic. Since the wild species are frequently used as folklore medicine, their phenology bears utmost importance from the point of their availability at different period of the year under the prevailing climatic condition (Upadhyaya & Saikia, 2007). Of course the users are very conscious about the life span of these species and people take care of them.

Scientific cultivation of five species has revealed that these species can establish well in the prevailing soil condition of the area and they can be grown for large scale production following appropriate technology that has to be developed for the purpose. In order to avoid the deleterious effects of synthetic chemical fertilizers and pesticides, organic agriculture is needed as an alternative to provide ecologically safe method of farming in this sector. Biopesticides of garlic and tulsi were found effective in controlling the Looper caterpillar that infested *A. vasica*. A final conclusion can be drawn after completion of successive phases of plantation and collection and compilation of data of their growth performance.

The present study revealed that although the users are conscious about the importance of these plants, reduction in number of MP species population is going on in the study area. Perhaps it is high time to take appropriate measures for conservation of this group of plants. The *ex-situ* strategy, although narrow, adopted at present seemed to be an effective means of conservation of this valuable group of plants in this area. At the same time emphasis should also be given for *in-situ* conservation. The practice of organic farming seems to be the best policy for conservation of this group of plant.

REFERENCES

Dahama A. K. 2005. *Organic Farming for Sustainable Agriculture*, Agrobios India, Jodhpur 342002, 276 pages.

Edwards C. A. 1988. Breakdown of animal, vegetable and industrial organic wastes by earthworms. *Agriculture, Ecosystems and Environment,* 24: 21-23.

Edwards C. A., Burrows I., Fletcher K. E. and Jones B. A. 1985. The use of earthworms for composting farm wastes. In: *Composting of Agricultural and Other Wastes* (Ed. J. K. R. Glasser), pp229-242.

Farooqi A.A. and Sreeramu B. S. 2004. *Cultivation of Medicinal and Aromatic Crops*. University Press (India) Pvt. Ltd., Hyderabad -500029.

Jain S. K. 1994. *Medicinal Plants*. National Book Trust, India.

Prajapati N.D., Purohit S. S., Sharma A. K. and Kumar T. 2006. *A Handbook of Medicinal Plants*. Agrobios India, Jodhpur-342001.

Pulikeshi M.B. and Amoji S. D. 2005. Assessment of compost and vermicomposts as biofertilizer through growth of food – crop seedlings. *Eco. Env. & Cons.,* 11(2):195-200.

Purohit S. S. and Vyas S. P. 2005. *Medicinal Plants Cultivation: A Scientific Approach,* Agrobios India, Jodhpur-342001.

Saikia L. R. and Upaduyaya S. 2006. Studies on the Bioresources and Indigenous Knowledge System of certain Ethnic groups of Eastern Himalayas., National Sem. on *Biodiversity & Indigenous Knowledge System*, Rajiv Gandhi Univ, Arunachal Pradesh, Oct 25 – 26.

Saravanan P., Elavarasan A., Annamalai A., Prasad R., Arumugam A., Kalaichelvan P. T. and Lalithakumari D. 2001. Agrotechnology for *Andrographis paniculata*. Souvenir, Nat. Sem. on *Com. Cult. Pross. and Mark. of Medicinal and Aromatic Plants*, JNKVV, Jabalpur (M.P.), India, page 83.

Sharma R. 2004. *Agrotechnology of Medicinal Plants*. Daya Pub. House, New Delhi-110035.

Smith R. L. 1986. In: *Medicinal Plants* by M. Agarwal, RBSA Publishers, Jaipur- 302003 (India).

Upadhyaya S. and Saikia L. R. 2007. Phenological observations of certain medicinal herbs of Dibrugarh, *Bull. Life Sci.*, XIII: 41-53.

13

Isolation of Endophytic Fungus from *Osbeckia stellata* Buch. Ham. Ex D.Don, A Medicinal Plant of the Pine Forests of Meghalaya, India

Ranjan K. Bhagobaty, Pramal Biswa and S. R. Joshi

1. Introduction

Areas having an exceptional concentration of endemic species and experiencing loss of habitat have been termed biodiversity hotspots (Myers *et al.* 2000). Together they occupy about 1.4% of the Earth's surface and contain about 44% of the world's total flora. Out of the 25 hotspots identified on the earth, two of them, namely Indo-Burma and the Western Ghats, are located in the Indian subcontinent (Groombridge1992; Myers *et al.* 2000). Due to the presence of a large number of primitive and ancient flowering plants in the region, the Indo-Burma region of Northeast India was considered as the cradle of ancient angiosperms (Takhtajan, 1969). Meghalaya, in Northeast India, which falls under the realms of the Indo-Burma biodiversity hotspot, is reported to have 3128 species of flowering plants including 1236 endemic species (Khan *et al.* 1997). The floristic richness of the state has also been recognised by several earlier workers (Hooker, 1854; Brandis ,1897; Kanjilal, 1934).

The local communities residing in the biodiversity-rich areas of the Northeastern Region of India have traditionally used and relied on herbs for treating various ailments (Kayang *et al.* 2005). Traditionally the leaves of *O. stellata* have been reported to be used for treating severe toothaches by the Apatani tribe of Arunachal Pradesh, India (Kala, 2005). The paste of the ripe fruits and leaves have been in use for time immemorial for treating cuts and wounds by the people of Nepal extending up to the Northeastern states of India (Balami, 2004). It has also been reported to be used as a piscicidal plant in Nepal and the adjoining areas by Chhetri & Joshi, 2002, and Joshi, 1988.

The World Health Organization (WHO) has listed 20,000 medicinal plants globally (Gupta *et al.* 1995); India's contribution (Singh, 2000) is15-20%. According to the WHO estimate, about 80% of the population in the developing countries depends directly on plants for its medicines (Pareek, 1996) (Mukhopadhyay, 1998). In India, about 2000 drugs used are of plant origin (Dikshit, 1999). Plant resources are depleting globally at an alarming rate and a number of economically and medicinally important plant species will soon be extinct. In the last few decades over-exploitation of forest resources has led to species loss. As a result, 20-25% of existing plant species in India has become endangered. Medicinal plants are now under great pressure due to their excessive collection or exploitation. The degree of threat to natural populations of medicinal plants has increased because more than 90% of medicinal plant raw material for herbal industries in India and also for export is drawn from natural habitat (Dhar *et al.* 2002).

The need to search and develop new, toxicologically safe and relatively cheaper drugs/ chemicals to treat diseases is ever growing, coupled with this is the diametrically opposite cry for *in-situ* conservation of medicinal plants in the few remaining undisturbed habitats. Drug resistance in bacteria, the appearance of life threatening viruses, the recurring problems with disease in persons with organ transplants and the tremendous increase in the incidence of fungal infections in the world's population have only underscored our inadequacy to cope with these medical problems. Added to this are enormous difficulties in raising enough food and protecting the existing crop plants against diseases and pathogens on certain areas of the earth to support local human populations. If alternate, easy to produce and cheap sources of bioactive natural compounds, having similar properties, as those derived from rare and endangered medicinal plants are found, then the anthropogenic pressure leading to over-exploitation of rare medicinal plants may be mitigated to a substantial extent. This point is exemplified by the case of the famous world's first billion dollar anti-cancer drug known as Taxol. Originally it is thought to be of pure plant origin and this compound was expensively extracted from the slow growing and relatively rare yew trees (*Taxus* species), since organic synthesis when having been accomplished, is not yet economically feasible (Nicolaou *et al.* 1994). After several years researchers found that the yew trees harbour in them a taxol-producing endophytic fungus, *Taxomyces andreanae* (Strobel *et al.* 1993; Stierle *et al.* 1993, 1995). Subsequently taxol-producing fungi were isolated from virtually all yew species and a myriad of trees in each of the major rain forests of the world (Li *et al.* 1996; Strobel *et al.* 1996). This discovery had immense value when we ponder upon the fact that a microbial source of the drug will eliminate the need to harvest and extract the slow growing rare yew trees for the drug. The exorbitant price of the drug would be significantly reduced, since taxol would conceivably be produced via fermentation in much the same way like most common antibiotics. Thus making this anti-cancer drug accessible to the common man in the economically poor nations of the world.

Thus, one probable approach to successfully tackle these interrelated and often conflicting demands/problems faced by mankind today is to search for novel compounds in the astronomic diversity of the yet untapped microbial diversity associated with plants of medicinal importance in the world's biodiversity hotspots.

It is noteworthy that, of the nearly 300,000 plant species that exist on the earth, each individual plant is a host to one or more endophytes. By definition, endophytes are those micro-organisms (mostly fungi and bacteria) that reside in the tissues beneath the epidermal cell layers and cause no apparent harm to the host (Stone *et al.* 2000). It is also hypothesized that some endophytes could have coevolved with their respective higher plant. Thus, closer biological associations may have developed between these endophytes in their respective hosts than is the case with epiphytes or soil-related organisms. The result of these associations may be the production of bioactive chemicals by the endophyte that are involved in the relationship. These novel natural products if isolated and characterized may have great promise for use in medicine, agriculture and industry. Only a few of these plants have ever been completely studied relative to their endophytic biology. These facts have prompted our search for novel endophytes and an examination of their secondary bioactive products in the traditionally used ethno-pharmacologically important plants of Meghalaya. The present study describes the isolation of an endophytic fungus from *Osbeckia stellata* Buch. Ham. ex D.Don, a medicinal plant of the Pine forests of Meghalaya, India.

2. Meterials and Methods

Collection of plant material

Osbeckia stellata was collected during the month of July 2007 from the *Pinus kesiya* forest undergrowth in the Umshing campus (Altitude 1500 m above M.S.L., 25°. 36' North latitude and 91°. 53' East longitude) of NEHU Shillong, Meghalaya (India). The specimens were identified by herbarium curator, department of Botany, NEHU, Shillong, and a sample specimen was deposited in the herbarium of the University.

Collection of soil samples

Composite soil samples were collected from the rhizosphere of healthy *O. stellata* plants. Samples were collected in sterilized polythene bags with the help of sterilized spatula as described by Arunachalam *et.al.* (1999).

Determination of soil pH

The pH of the supernatant was measured by the method as described by John *et al.* (2002).

* As per Google Earth 2007 statistics.

Enumeration of microorganisms

Enumeration of microorganisms from soil was carried out as described by Arunachalam *et. al.* (1999)

Isolation of endophytic fungus from *O. stellata*

Surface sterilized root and stem of *O. stellata* were air dried and flamed before removing the outer layers. Two centimeters long pieces of these roots and stems were placed on Petri plates containing water and agar plates as described by Strobel *et al.* 1996 and incubated at 37 °C. After incubation for several days, hyphal tips of developing fungi were aseptically removed and placed on potato dextrose agar (PDA) as per the procedure described by Strobel *et al.* (2005).

Fungal Growth and Storage

The fungus was subsequently grown on PDA plates and also in liquid PD medium. Pure culturing of the fungus was done on 14 days interval from the working culture plates. Agar blocks containing the fungus were placed in sterile distilled water and stored at 4° C. Fungal viability was determined as per the protocol described by Strobel *et al.* (2005).

Morphological characterization of the endophytic fungus

The fungus was stained using the aniline blue as per the standard protocol described by Cappucino & Sherman (1992) and observed under a Motic phase contrast trinocular research microscope model BA-450PH (Feintechnik, Germany).

3. Results and Discussion

The population of various microbial communities in the rhizospheric soil of *Osbeckia stellata* Buch. Ham. ex D. Don is shown in Table 1. Present study indicated that the total bacterial population in the rhizospheric soil of *O.stellata* was $3.52X\ 10^6$ CFU/ gm of wet wt. of soil. Total actinomycetes count was $3.4X10^4$ CFU/gm wet wt. of soil; whereas the total soil fungal population was determined to be $6.0\ X\ 10^2$ CFU/gm wet wt. of soil.

Table 1

Enumeration of microorganisms from Rhizospheric soil of *Osbeckia stellata*

Parameters	O. stellata rhizospheric soil sample
PH	5.9
Bacteria	$3.52X\ 10^6 \pm 0.02 \times 10^6$
Actinomycetes	$3.4X10^4 \pm 0.18 \times 10^4$
Fungi	$6.0\ X\ 10^2 \pm 0.21 \times 10^2$

Values are mean± SD in cfu/g of wet weight of soil (n= 3X2).

Arunachalam *et al.* (1999) while studying the effect of soil properties on microbial populations in the *Pinus kesiya* forests of Meghalaya has reported the number of fungal propagules to be significantly higher in forested ecosystems than in the non-forested ecosystems. He further observed that in the forest re-growths of three different ages, the populations of both bacteria and fungi showed a gradual increase with the age of the stand. He reported that the bacterial population in 16-year-old forest re-growths to be $1.90X10^4$ CFU/ gm dry soil and fungi to be $2.00X10^3$ CFU/ gm dry soil. The values that we have obtained from the study of the rhizospheric soil of *O.stellata* are comparable with these figures and also give an indication of the rhizospheric effect of microbial growth promotion in and around the plant roots. This is primarily due to the release of growth factors by the plant into the soil through its roots. The microbial population in the vicinity of its roots is therefore comparatively higher than the soils where the plant roots are absent.

The pH of the rhizospheric soil of *Osbeckia stellata* Buch.Ham. ex D.Don was 5.90 (Table 1). The pH of the soil found here is comparable to that reported by Arunachalam *et al.* (1999) and John *et al.* (2002). They also reported that the pH of the soil under *Pinus kesiya* forests in Meghalaya ranges from 5.2 to 5.4, indicating that the soil under *Pinus kesiya* stands to be acidic.

The colony, growth and morphological characteristics of the endophytic fungi are presented in Table 2. White cottony fungal mycelia were observed to emerge out of the plant tissues after four days of incubation. The fungal colony fully covered the water agar plates after 6-7 days of incubation and produced a characteristic fruity odour, growth of other fungi was not observed in the plates. Similar odour characteristics were also reported by Woropong *et al.* 2001 for the endophytic fungus *Muscodor albus* isolated from *Cinnamomum zeylanicum*. *Muscodor albus*, which literally means 'stinky white'

(Strobel, G. 2006), was later shown to produce bioactive volatile organic compounds (VOCs) with potent antibiotic activity against a broad range of plant and animal pathogens. (Strobel *et al.* 2001). It is therefore likely that the endophytic fungus isolated from *O. stellata* also produces volatile anti-microbial compounds. It is also likely that due to the antimicrobial compounds produced by the fungus, it was able to prevent other microorganisms from growing in the water agar plates even after 15 days of incubation.

Table 2
Characteristics of the fungal isolate

Sl. No.	Characteristics	Description
1	Colony colour	White
2	Colony texture	Cottony
3	Colony shape	Round
4	Hyphae	Aseptate
5	Sporulation	After 15 days of incubation in PDA medium at 37 ° C
6	Spore colour	Greenish brown

The growth of the fungal colonies on PDA plates could be visualized in 2-3 days after inoculation at 37°C. The rapid growth of the fungi when transferred to PDA is justified, since it has starch and dextrose as a source of carbon to meet its metabolic, whereas in the case of water agar the constituents were only water and agar. Agar being used to provide a solid surface for the growth of the fungus. The nutrients for growth and development of the fungi, in the case of water agar plates being obtained from the plant material itself.

Upon comparison of the colony, growth and morphological characteristics of the isolated endophytic fungi with those of well defined fungal species as described in the reference book 'Fungi in Agricultural Soils' (Domsch and Gams, 1972) the fungus was found to have similar characteristics with *Mortierella hyalina* (Harz) W.Gams.

M. hyalina has a worldwide distribution in the temperate zone. It has been reported from North Scotland down to Central Europe and also from New Zealand. It is quite frequent in forests soils particularly under pine and grasslands. It has been more frequently observed in organic than inorganic particles. It primarily occurs in the upper soil horizons, but has also been reported from depths of up to 135 cm from the soil surface. *M. hyalina* has also been reported to be isolated from the roots of *Trifolium repens* (Domsch and Gams, 1972).

Further, molecular characterization with 18S rDNA and ITS sequence specific primers of the fungal strain will reveal the phylogenetic relationship with other related species with help us ascertain its identity in totality.

4. Conclusion and Future Perspectives

To our knowledge this is the first report of isolation of an endophytic fungus from *Osbeckia stellata* Buch. Ham. ex D.Don from the pine forests of Meghalaya. Efforts are currently on to characterize the metabolite(s) produced by this endophytic fungi using OSMAC (one strain many compounds) approach as described by Bode *et al.* 2002, which will involve utilization of techniques like thin layer chromatography, gas chromatography and mass spectroscopy. These metabolites can be tested for their efficacy against pathogenic microorganisms and as anti-hyperglycemic agents. The metabolite(s) characterized from the endophytes may have a correlation with the active principles produced in the host plant as this may be the result of the interaction of the host with the endophytes.

Figure 1 (a). *Osbeckia stellata*

Figure 1(b). Flowers of *O.stellata*

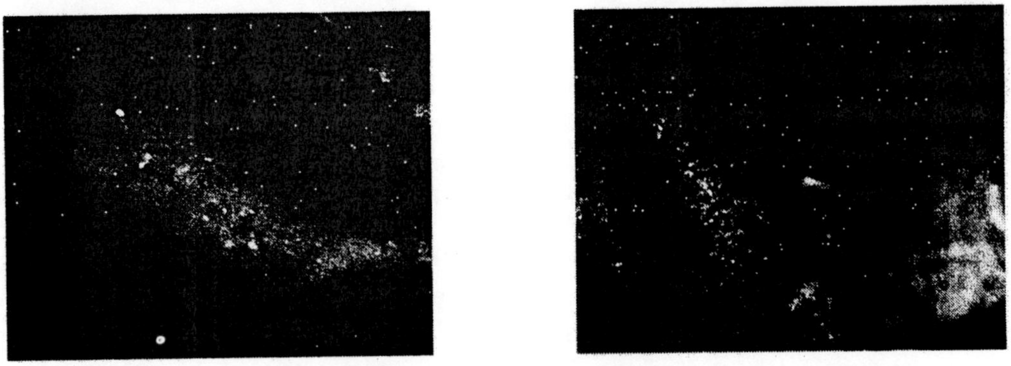

Figure 2 (a) & (b). Fungal hypahe emerging out of the plant materials (4 X magn.)

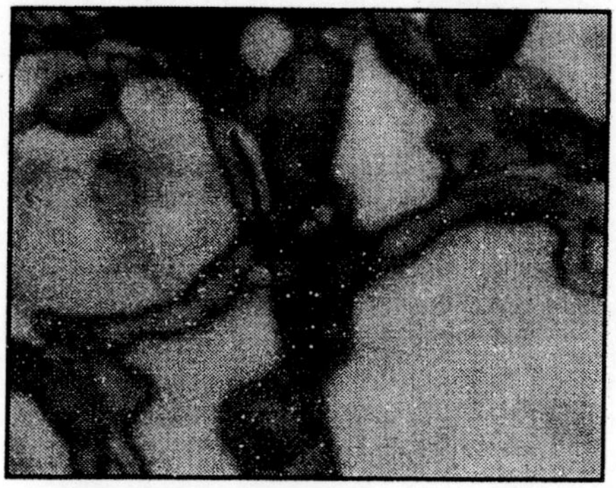

Figure 3. Fungal hyphal structure (40 X magn.)

Acknowledgements: The authors wish to acknowledge Dr. (Mrs.) R.N.K. Hooroo, Department of Zoology, NEHU, Shillong, for the help provided during digital photography of the fungal samples and are highly thankful to herbarium curator Dr. P.B. Gurung, Department of Botany, NEHU, Shillong, for identifying the collected plant specimens.

REFERENCES

Arunachalam K., Arunachalam A. and Melkania N.P. 1999. Influence of soil properties on microbial populations, activity and biomass in humid subtropical mountainous ecosystems of India. *Biol Fertil Soils*. 30: 217–223

Balami N. P. 2004. Ethnomedicinal uses of plants among the newar community of pharping village of Kathmandu district, Nepal. *Tribhuvan Univ. Jour.* 24(1): 13-19

Bode H. B., Bethe B., Hofs R. and Zeek A. 2002. Big Effects from Small Changes: Possible Ways to Explore Nature's Chemical Diversity. *ChemBioChem*, 3: 619-627.

Brandis D. 1897. Indian Forestry Working. Oriental Institute, Woking, India.

Cappucino J. G. and Sherman N. 1992. *In Microbiology: A Laboratory Manual*. Addison-Wesley Pub Co, USA.

Chhetri R.B. and Joshi R. 2002. Ethnobotany of non-timber forest product in Langtang National Park, Nepal. *Proceedings of International Seminar on Mountains* – Katmandu. RONAST, Katmandu, Nepal. pp 469-472.

Dhar U., Manjkhola S., Joshi M., Bhatt A., Bisht A. K. and Joshi M. 2002. Current status and future strategy for development of medicinal plants sector in Uttaranchal, India. *Curr. Sci.*, 83: 956–964

Dikshit V. K. 1999. Export of medicinal plants from India: need for resource management. In *Biodiversity – North-East India Perspectives: People's Participation in Biodiversity Conservation* (Eds Kharbuli B., Syiem D. and Kayang H.), NEBRC, North-Eastern Hill University, Shillong, pp. 85–88.

Domsch and Gams. 1972. *In Fungi in Agricultural Soils*. Halsted Press Division, New York, USA.

Groombridge B. 1992. *Global Biodiversity Strategy*. World Conservation and Monitoring Centre, Chapman & Hall, London.

Gupta R. and Chadha K. L. 1995. Medicinal and aromatic plants in India. *In Advances in Horticulture, Medicinal and Aromatic Plants* (Eds Chadha, K. L. and Gupta, R.), Malhotra Publishing House, New Delhi, pp. 1–44.

Hooker J.D. 1854. *Himalayan Journals*, Vols. I & II. Today & Tomorrow Printers & Publishers, New Delhi, India.

John B., Pandey H.N. and Tripathi R.S. 2002. Decomposition of fine roots of *Pinus kesiya* and turnover of organic matter, N and P of coarse and fine pine roots and herbaceous roots and rhizomes in subtropical pine forest stands of different ages. *Biol Fertil Soils* .35: 238–246

Joshi A.R. 1988. A Study of the Environmental Relationship of Certain Village Communities in the Central Development Region of Nepal, Ph.D. Thesis, University of Wales, Cardiff, UK.

Kala C. P. 2005. Ethnomedicinal botany of the Apatani in the Eastern Himalayan region of India. *Jour of Ethnobiol and Ethnomed* 1:11(http://www.ethnobiomed.com/content/1/1/11)

Kanjilal U.N. 1934. A brief ecological sketch of the botany of Assam. *Flora of Assam*, Vol. 1(a). Government Press, Shillong, India.

Kayang H., Kharbuli B., Myrboh B. and Syiem D. 2005. Medicinal plants of Meghalaya. In: Bioprospecting & Ethnopharmacology. *Acta Horticulturae* 675: 75–80.

Khan M.L., Menon S. and Bawa K.S. 1997. Effectiveness of the protected area network in biodiversity conservation, a case study of Meghalaya State. *Biodiversity and Conservation* 6: 853–868.

Li J.Y., Strobel G.A., Sidhu R., Hess W.M. and Ford E. 1996. Endophytic taxol producing fungi from Bald Cypress *Taxodium distichum*. *Microbiology*.142: 2223–2226.

Mukhopadhyay S. 1998. Conservation, protection and biodiversity of medicinal plants. In *Prospects of Medicinal Plants* (Eds Gautam,P.L. et al.), Indian Society for Plant Genetic Resources, New Delhi, pp. 15–28.

Myers N., Mittermeir R.A., Mittermeir C.G., Da Fonseca G.A.B. and Kents J. 2000. Biodiversity hotspots for conservation priorities. *Nature* 403: 853–858.

Nicolaou K.C., Yang Z., Liu J.J., Ueno H., Nantermet P.G., Guy R.K., Claiborne C.F., Renaud J., Couladouros E.A., Paulvannan K. and Sorensen E.J. 1994. Total synthesis of taxol. *Nature* (Lond.), 367: 630–634.

Pareek S. K. 1996. Medicinal plants in India: Present status and future prospects. *In Prospects of Medicinal Plants* (Eds Gautam, P. L. *et al.*), Indian Society for Plant Genetic Resources, NBPGR Campus, New Delhi, pp. 5–14.

Singh H. B. 2000. Alternative source for some conventional drug plants of India. *In Ethnobotany and Medicinal Plants of Indian Subcontinents* (Ed. Maheshwari, J. K.), Scientific Publishers, Jodhpur, India, pp. 63–78.

Stierle A., Stierle D. and Strobel G.A. 1995. The search for taxol producing microorganisms among the endophytic fungi of the Pacific yew, *Taxus brevifolia. J. Nat. Prod.* 58: 1315.

Stierle A., Strobel G.A. and Stierle D. 1993. Taxol and taxane production by *Taxomyces andreanae. Science* (Wash., D.C.), 260: 214–216.

Stone J. K., Bacon C. W. and White J.F. 2000. An overview of endophytic microbes: endophytism defined. In Microbial Endophytes, pp. 3-29. Eds by C. W. Bacon & J. F. White. Marcel Dekker. New York:

Strobel G.A. and Daisy B. 2003. Bioprospecting for Microbial Endophytes and Their Natural Products Microbiology and molecular biology reviews 67(4): 491–502

Strobel G.A., Manker D.C. and Mercierv J. 2005. *Endophytic Fungi and Methods of Use.* (http://www.freepatentsonline.com/6911338.html)

Strobel GA., Dirksie E, Sears J. and Markworth C. 2001. Volatile antimicrobials from a novel endophytic fungus. *Microbiol*, 147: 2943-2950

Strobel G., Yang X., Sears J., Kramer R., Sidhu R.S. and Hess W.M. 1996. Taxol from *Pestalotiopsis microspora*, anendophytic fungus of *Taxus wallichiana. Microbiology*. 142: 435–440.

Strobel G.A., Stierle A., Stierle D. and Hess W.M. 1993.*Taxomyces andreanae* a proposed new taxon for a bulbilliferous hyphomycete associated with Pacific yew. *Mycotaxon*, 47: 71–78.

Takhtajan A. 1969. *Flowering Plants: Origin and Dispersal.* (Eds Singh B, Singh M P and Koeltz O). Science Publishers, Dehra Dun, India.

Woropong J., Strobel G.A., Ford E.J., Li J.Y., Baird G. and Hess W.M. 2001.*Muscodor albus* anam. nov. An endophyte from *Cinnamomum zeylanicum*. *Mycotaxon*, 79:67-79.

14

A Study on Medicinal Values of Some Species of Family Polygonaecae

Alee Sarma and *Manashi Kalita*

1. Introduction

Medicinal plants are the local heritage with global importance. The world is endowed with a rich wealth of medicinal plants. Herbs have always been the principal forms of medicine in India and presently are becoming popular throughout the developed world as people strive to stay healthy in the face of chronic stress and pollution. These plants also play an important role in the lives of rural people, particularly in remote parts of developing countries with few health facilities. Medicinal plants have curative properties due to the presence of various complex chemical substances of different composition known as secondary plant metabolites. The plant metabolites according to their composition are grouped as alkaloids, glycosides, corticosteroids and essential oils.

The entire Northeastern region is rich in bioresource especially in terms of its forest wealth, biodiversity, varied environment and has also been classified as ecological "Hot Spot". The region has plenty of medicinal and aromatic plants which have been used by people for various medicinal and other useful purposes. The region could have a very high value addition which results in higher employment generation and increased income if propagation and cultivation of high valued medicinal and aromatic plants is adopted.

The paper emphasized on medicinal uses of some species of Polygonaceae. They are herbs, shrubs or rarely trees comprising about 30 genera and 1,000 species. The leaves are simple, alternate or rarely opposite and usually stipulate. The membranous, sheathing stipule is called an ocrea. The flowers are actinomorphic, usually small and are bisexual or occasionally unisexual. The perianth is usually in two undifferentiated whorls comprising 3-6 distinct or basally connate sepals. The androecium consists of 6-9, commonly 8 distinct or basally connate stamens, usually in two whorls. The gynoecium consists of a compound pistil of usually 3 carpels, an equal number of style branches or styles, and a superior ovary with one locule containing a single basal ovule. The fruit is usually a 3-angled or 3-winged achene.

In this work the broad outline system of classification of Benthan and Hooker's (1883) was adopted. But the generic limits have been accepted as redefined, especially that of *Polygonum* L. (SI). Nomenclature is based mainly on Gross (I. c.), Nakai (1952), Headberg (I.c.) and Haraldson (I.c.). The generic delimitation of *Polygonum* L. (SI) to several genera like *Polygonum, Bistorta, Persicaria, Koenigia, Aconogonon.*

The information collected during the survey covering about 35 species belonging to seven genera of the family polygonaceae. Some of the uses (both edible and medicinal) of this family are given below:

2. Results

1. *Bistorta major S.F. Gray*

Chemical constituent: 100gms of leaves contains 3g of protein, 0.8g fat, 7.9g of Carbohydrate and 3.2g of fibre.

Edible uses:

Leaves – Use raw or cooked. When the leaves are young they have a fairly mild flavour. They make an excellent substitute for spinach. The leaves are good sources of Vitamin A & C.

Seed - Raw or cooked.

Root - Used raw or cooked. It is rich in starch and tannin, it is steeped in water and then roasted in order to reduce the tannin content. It is a tasty and nutritious food. The root is also boiled and used in soup and stews; it can be dried and ground into powder for making bread. The root contains 30% starch, 1% calcium oxalate and 15 to 30% tannin.

Medicinal uses: Bistort is one of the strongest astringent of all herbs and it is used to contract tissues and reduce blood flow. The root is a powerful astringent, demulcent, diuretic, laxative and strongly styptic. It is much used in the treatment of external bleeding, diarrhea, dysentery, cholera etc. It is also taken internally in the treatment of catarrh, irritable bowel syndrome, peptic ulcers and excessive menstruation. Externally, it makes a good wash for small burns and wounds and also used to treat pharyngitis, stomatitis, vaginal discharge etc. A mouth wash or gargle is used to treat spongy gums, mouth ulcers and soar throats.

Other uses: Tannins can be extracted because the root contains up to 21% tannin.

2. *Fagopyrum dibotrys* (D. Don) Hara

Local Name : Hanjareng (karbi) , Dhemsi-sak(as)

Chemical constituent: The plant contains antioxidant phenolics, diboside and flavonol glycoside.

Edible uses:

Leaves – Used raw or cooked. The leaves are rich in rutin and they make a healthy addition to the diet.

Seed - It can be sprouted and eaten raw or cooked and also used as a cereal. Seeds were dried and ground into powder that serves as a thickening agent in soups. The seed is rich in vitamin B_6.

Medicinal uses

The whole plant is anodyne, antihelmintic, antiphlogistic, carminative, depurative and febrifuge. It stimulates blood circulation. A decoction is used in treatment of traumatic injuries, menstrual irregularities, snake and insect bites, dysmenorrhoea, inflammation and lumbago. The leaves are rich in rutin which is a capillary tonic, anti-odemic, anti-inflammatory, antispasmodic etc. Rutin also inhibits carcinogenesis and protects against radiation.

3. *Fagopyrum esculantum* Moench (Buckwheat)

Local Name : Sutia-lafa(as)

Composition: Protein content varies from 13-15 % of the groats. The main protein fraction is globulin, which represents almost half of all proteins. A characteristic feature of buckwheat protein is a very low content of prolamin.

Starch is the major carbohydrate in buckwheat. Starch granules in the endosperm are polygonal or round in shape with diameter ranging from 2 to 12 ìm. The majority of granules are 6-8 ìm in diameter. The dietary fibre content may vary from 5-11%. In the whole buckwheat grain the total lipid content ranges from 1.5 - 4%. The highest content is in the embryo (9.6-19.7%); the endosperm contains 2 - 3% and the hulls 0.4 - 0.7 %. Buckwheat oil contains 16-20% saturated fatty acids, 30-45% oleic acid and 31-41% of linoleic acid. Palmitic, oleic, linoleic and linolenic acids account for about 95% of buckwheat total fatty acids content. The ash content of buckwheat ranges from 2 - 2.2%, depending upon the variety. The polyphenolic compounds of buckwheat were determined at 0.7% in the hulls and 0.8% in the groats. Buckwheat flour contains various kinds of vitamins, such as B_1, B_2 and niacin at relatively high levels (Pomeranz, 1983).

Edible uses:

Leaves – Used raw or cooked. The leaves are rich in rutin that are healthy addition to the diet.

Seed - Used raw or cooked. The sprouted seeds can be used in salads. It can also be ground into a powder and used as a cereal; when it can be made into pan-cakes, noodles, breads or be used as a thickening agent in soups. It is rich in vitamin B_6. An excellent beer can be brewed from its grains.

Medicinal uses: Leaves are good source of rutin which are useful in treatment of a wide range of circulatory problems; it dilates the blood vessels, reduces capillary permeability and reduces blood pressure. Leaves and shoots are internally used in the treatment of high blood pressure, gout, varicose veins, radiation damage etc. It often combines with lime flower to treat hemorrhage of retina. A poltice made from the seeds has been used for restoring flow of milk in nursing mothers. An infusion of the herb has been used in the treatment of skin disease.

Other uses: A very good manure plant. It can be used to reclaim badly degraded soils. A blue dye is obtained from flowers.

4. *Persicaria barbata* (L)Haras

Local Name: Atalari(as), Bekhunjubag(beng)

Edible uses: Tender leaves and shoots- cooked as vegetable. They have a pungent flavour.

Medicinal uses: Seeds are used to relieve the gripping pain of colic. A decoction of leaves and stalk is used as a stimulating wash for ulcers. The root is astringent and cooling. A paste of root is used externally in the treatment of scabies.

5. *Persicaria chinensis* (L) H. Gross

Local name : Madhuri tenga(as)

Edible uses: The young shoots are used as curries and fodder for cattle. **Medicinal uses:** The whole plant is used medicinally in general debility.

6. *Persicaria glabra* (Willd.) M. Gomes

Edible uses: Leaf is used as vegetables. **Medicinal uses:** Leaf infusion is used in fever and in colic pain. The plant extract is used in pneumonia and root stalk is used in piles, jaundice and in general debility.

7. ***Persicaria hydropiper*** **(L) Spach**

Local Name : Bon-jaluk(as),Water pepper(eng);

Chemical composition: Protein = 7.54%. Fat = 1.86%. Carbohydrate = 7.99%.

Edible uses: Leaves and stems are eaten raw or cooked. Seeds used as pepper substitute. The sprouting seeds or young seedlings can be used as a garnish or additive to salads.

Medicinal uses: Due to its astringent properties it is useful in treating - bleeding, skin problems and diarrhea. Leaves contain polygonic acid which promotes the coagulation of blood; polygonone- containing ethereal oil which lowers blood pressure (Chopra *et al.* Glossary of Indian Medicinal plants, 1956).The leaves are also anti-inflammatory, astringent and stomachic. The seed is carminative, diuretic and stimulant. The whole plant, either on its own or mixed with other herbs, is decocted and used in the treatment of a wide range of ailments including diarrhea, dyspepsia, itching skin, excessive menstrual bleeding, swollen and inflamed areas. Roots are used to expel worms. It is effective as a cure for toothaches, epilepsy, gangrene and gout (James Dwyer *et al.* Magic and medicine of plants - Readers Digest, 1986). Animal studies suggest that the plant has contraceptive effects.

Other uses: The plant yields a golden yellow dye and an essential oil (S.P.Ambast, The use plants of India, 1986).

8. ***Persicaria lapathifolia*** **(L.) S.F. Gray**

Chemical constituent: It contains a number of acylated kaempferol glycosides like kaempferol, galactopyranoside, quercetin and rhamnopyranoside.

Edible uses: Young leaves and seed used raw or cooked.

Medicinal uses: The whole plant is antiseptic and astringent. It is used in the treatment of stomach complaints and fevers. The plant produces a soft white mass, froth like that of soap. It is applied externally to burns.

Other uses: The soft white mass, froth like that of soap is used for bathing and washing clothes.

9. ***Persicaria orientalis*** **(L.) Spach**

Local name : Bora-panimirch (beng)

Medicinal values: The plant is a good tonic and vulnerary. The flowers are said to thinning the blood, remove obstructions and ease pain. The leafy stems are used in the treatment of hernias. A decoction of the ripe fruits is used in the treatment of hepatitis, sloughing ulcers, tympanites and cancer. The seed is said to relieve flatulence, fever, thirst, brighten the eyes and benefit the breath.

Edible uses: Young shoots and seeds – are cooked. It is a standard vegetable in Indo-China.

Other uses: Concentrated infusion of the plant is used to poison fish.

10. *Persicaria perfoliata* (L.) H. Gross

Local Name : Baghachora

Edible uses : Tender young leaves and shoots – are used raw or cooked as vegetables. The ripe seeds are eaten fresh, especially by children.

Medicinal uses: The whole plant is depurative, diuretic and febrifuge. It is also used to stimulate blood circulation. A decoction is used in the treatment of dysentery, boils and abscesses, poisonous snake bites, haematuria, urinary problems and traumatic injuries. The juice of the leaves is used in the treatment of back-aches.

11. *Persicaria micrecephala* (D. Don) H. Gross

Chemical Constituent: Minerals such as sulphur, iron and chlorine are present.

Edible uses: Leaves are acidie. Young tops are used in curries and chutnies. Seeds are used as condiments.

12. *Polygonum aviculare* L.

Local Name : Indrani (as)

Chemical constituent :100gms of leaves contains: Protein 1.9g; Fat 0.3g; Carbohydrate 10.2g and fibre 3.5g.

Edible uses: The plant is used as vegetable and as fodder for cattle. It is also used as pot herb, seeds either whole or dried are ground into a powder for use in pancakes and biscuits. Leaves are used as tea substitute.

Medicinal uses: It is a safe and effective astringent and diuretic herb that is used mainly in the treatment of dysentery and haemorrhoids. It is also taken in the treatment of pulmonary complaints because the silicic acid it contains strengthens connective tissues in the lungs. It was formerly used as an astringent both internally and externally in the treatment of wounds, bleeding, piles and diarrhea. It is useful in removing stones. It is an excellent remedy to stop bleeding of the nose and to treat sores.

Other uses: The leaves yield a blue dye. Yellow and green dyes are also obtained from the whole plant.

13. *Polygonum plebeium* R. Br.

Local Name : pani-jaluk (as)

Medicinal uses : The roots are used in bowel complaints. Dried and powdered plants are taken internally in pneumonia.

14. *Polygonum dumetorum* L.

Edible uses : Leaves and seeds – are used raw or cooked.

Medicinal uses: It is a purgative.

15. *Polygonum molle* D. Don

Chemical constituent: It has got high content of oxalic acid giving it a lemon taste.

Edible uses: young shoots are used raw or cooked as vegetable. Pleasantly acidic in nature, they are used like rhubarb.

Medicinal uses: The whole plant is astringent.

16. *Rumex maritimus* L.

Local name : Bon-palang (as)

Chemical constituent : The plant contains n-dotriacont-3-one, named as Maritimone, n-heptacos-18,20,22-trien-1-ol,named as Rumexenol, 22-n- hexyl-12-(a-tetradecahydroanthracenyl)-ndecan-3-ol-1-oic acid, named as Rumexoic acid, quercetin, sitosterol glycoside, 3,7,11,15-tetramethyl-n-octadac-13-en-3,4,6,7,8,11-hexol-12-one, named as Rumexenone, 8-normethyl-16-(19,22-dimethyl-19,22-dihydroxy-heptanyl)-spongiene-5-en-7-on-8-ol-22-b-glucopyranosyl-β-D-glucopyranoside, named as Maritiaspongiene. Different extracts of these plants exhibited significant anti-inflammatory and anti- bacterial activity against *Staphylococcus aureus* (G+).

Edible uses: Leaves and seeds are cooked before used. Seeds can be ground into a powder then mixed with water and cooked like porridge.

Medicinal uses: The leaves are applied externally to burns. The seeds are Aphrodisiac. They contain about 5% tannin.

Other use : Dark green to brown and dark gray dyes can be obtained from the roots of many species in this genus.

17. *Rumex nepalensis* Spreng.

Local name : Tor-boura

Chemical constituent: The antibacterial property of *Rumex nepalensis* Spreng. was evaluated against some strains of bacteria. The methanol extract of the roots (tested at 200-1000 micro g/disc) showed significant result for antibacterial activity.

Edible uses: Tender young leaves and shoots are cooked as vegetable.

Medicinal uses: The root is purgative. It is used as a substitute for rhubarb. A strong decoction of the root is applied in dislocated bones. A part of the root is applied to swollen gums. The leaves are used in the treatment of colic. The juice of the leaves is applied externally to relieve headaches. A decoction of the plant is used to wash the body in order to alleviate body pain.

18. *Rumex vesicarius* L.

Local name : suka-sak (as)

Chemical constituent: The chemical composition of the leaves indicates that it is a good source of minerals, a moderate source of protein and ascorbic acid; it is high in oxalic acid and low in lipids and tocopherol.

Edible uses: Leaves used as vegetable. Leaves are acidic, they are used like sorrel as a flavouring in salads or as spinach.

Medicinal uses: They are aperients, astringent, diuretic and cooling. They are roasted and used in the treatment of dysentery. They are useful in treating stomach problems, to allay the pain of toothache and to check nausea.

3. Conclusion

It has been observed that a number of species of the family Polygonaceae have medicinal values and are widely available in Assam. The people of the state have been using these herbs for ages in treatment of various diseases. Besides, they are also used as nutritional food since they are rich sources of minerals. Though the medicinal properties of these herbs are well known, mineral content of the herbs are not well known. These can be used for prevention of various diseases caused by mineral deficiency. As the climate is suitable for growth of these herbs in the state and their demand is very high in the national and international market, proper plantation of these herbs can be undertaken scientifically. It is to be noted that most of these species are treated as weeds due to lack of knowledge about their medicinal values and mineral content.

Pictures of Some of the species of family Polygonaceae

Bistorta major

Fagopyrum esculentum

Fagopyrum dibotrys

Persicaria barbata

Persicaria hydropiper

Persicaria chinensis

Persicaria lapathifoli

Persicaria orientalis

Persicaria perfoliata

Persicaria micrecephala

Polygonum aviculare

Polygonum plebeium

Polygonum molle

Rumex maritimus L.

Rumex nepalensis

Rumex vesicarius

REFERENCES

Agarwal V.S. 1986. *Economic Plants of India*, Dehradun.

Ahluwalia K.S. 1952. Medicinal Plants of Kangra, *Ind. For.*

Ambast S.P. 1986. *The Useful Plants of India*, CSIR, N. Delhi.

Anonymous 1942-1976. *Wealth of India*, Vol. I-II, Coun. Sci. Industries, N. Delhi.

Anonymous 2004. *Assam Forest at a Glance*, Published by Chief Conservator of Forests, Assam.

Anonymous 1998. *Economic Survey of Assam*, Published by Govt. of Assam.

Asolkar L. and Kakkar K.K. 1992. *Second Supplement of Glossary.*

Bakharu K.K. 1992. *Herbs That Heal*, N. Delhi.

Barthakur S. and Jain. 1980. *Ethnobotany of the Mikirs of India.*

Bentley R. and Henry T. 1980. *Medicinal Plants*, Vol. I-IV, London.

Chiej R. 1984 *Encyclopaedia of Medicinal Plants.* MacDonald.

Chopra R.N. and Naiar S.L. 1956. *Glossary of Indian Medicinal Plants*, & Chopra, I.C. CSIR, N. Delhi.

Clarke O.J. 1965-1981. *Indian Medicinal Plants with Active Principles*, Part-I (A-K) CSIR, N. Delhi.

Dey A.C. 1980. *Indian Medicinal Plants Used in Ayurvedic Preparation*, Dehradun.

Jain, S.K. 1968. *Medicinal Plants*, National Book Trust, N. Delhi.

Journal of Food Composition and Analysis , Volume 19, Issues 6-7, September-November 2006, Pages 552-555

Kirtkar K.R. and Basu B.D. 1980-1984. Indian Medicinal Plants, Vol. I-IV, Alahabad

'Phytochemical and Pharmacogistical studies on Some Indian Medicinal Plants.' PhD thesis of Syed Tarique Abdullah, Jamia Hamdard University, 2002

Phytother Res, 2003. May, 17(5), 558 – 9

Pomeranz Y. 1983. Buckwheat: structure, composition and utilization. *CRC Crit. Rev. Food Chem.* 19:213-258

Srivastava R. C. 1989. *Drug Plant Resources of Central India*, Today & Tomorrow's Printers & Publishers, N. Delhi.

15

Traditional Knowledge System of Medicines:
A Case Study in Khasi Hills

Mandy M. Kharbyngar, Kyntiewlang M. Kharbyngar and Jacqueline Dkhar

1. Introduction

The principal objects of Herbal Medicines are not only to cure human beings from diseases but to prevent them from all kinds of miseries in life. A medicinal herb deals with entire nature of life. Happy life is the union of the body, senses, mind and spirit. Therefore, man is a combination of mental, physical and spiritual factors.

Medicinal herbs/plants in Khasi Hills especially in rural area are a gift of nature. Medicinal herbs are (1) Perennial where medicines can be prepared from bark of tree, leaves, roots, stem, fruit, seed etc. (2) Seasonal where medicine is getting from leaves, roots, stems, fruits, flowers etc. A local tradition used different parts among several medicinal plants to cure specific ailments has been in vogue from olden days. Our country has wide variation in climate, soil, altitudes and latitudes. Nature has bestowed on us a very rich and large number of diverse types of plants growing wild in different parts of the country. Increasing awareness programmes on the potential activities of these interesting and useful plants have encouraged many innovative and progressive growers and entrepreneurs to take up their cultivation apart from health care; this provides means of livelihood to scores of people.

However, our knowledge of medicinal plants has mostly been inherited traditionally. The use of plants for curing various ailments are not confined to the qualified doctors only but is known to several households as well. There are many interesting and sometimes astonishing things to learn from herbal practitioners. Spreading and preserving this knowledge on medicinal plants and their uses have become important for human existence. There has been a growing tendency all over the world to shift from synthetic to natural based product including medicinal plants.

In modern medicines also, plants occupy a very significant place as raw materials for some important drugs, although synthetic drugs and antibiotics brought about a revolution in controlling different diseases. But these synthetic drugs are out of reach to million of people especially those who lives in remote places, thus, they are sole dependants on traditional healers whom they know and trust.

2. Some Common Plants used in Traditional Knowledgement

1. Okra

Botanical Name: *Abelmoschus esculentus*

Medicinal uses

1. Soothes and protects the alimentary tract.
2. Relieves flatulence and gripping pains from the stomach.
3. Used in case of pneumonia, urinary problem, heart diseases and diabetes.

Methods –The leaves are dried, grind to powder. 1 tsp. of powder mixed in 2 tsp hot water, cool and drink 3 times daily in case of pneumonia. Dry root powder helps in flatulence and stomach pain. Fresh fruit is cut longitudinal into half and kept in a glass of boil cool water overnight. It helps in diabetes. Seeds are use for heart disease.

2. Onion (Kh – Piat)

Botanical name: *Allium cepa*

Medicinal uses

1. Used as stimulant
2. Promotes the flow of urine
3. Promotes menstrual flow
4. Soothes and protects the alimentary tract and relieves inflammation
5. Purify the blood

6. Use against tuberculosis, typhoid, fever, cough, chest and lung problems and skin diseases
7. Good for insect bites, boil containing pus, nasal bleeding and arthritis.

Methods- Onion eaten raw is good for the various disorders and diseases mentioned above. Grind onion mixed with little pure honey is externally used for insect bites, boil containing pus, nasal bleeding and arthritis.

3. Garlic (Kh – Rynsun)

Botanical name: *Allium sativum*

Medicinal uses

1. Used to kill or expel worms from the body
2. Stimulates flow of saliva. help in digestion, relieve flatulence and gripping pains from stomach
3. Promotes the flow of urine.
4. Promotes the removal of mucous secretions.
5. Used to reduce fever.
6. Can be used in chronic bronchitis, tuberculosis, pneumonia, asthma and other respiratory ailments.
7. It is good for colds and can be used in sore throat, hoarseness and tonsillitis.
8. Good for hypertension, skin problems and prevention of grey hair.

Methods - Use of raw garlic is good for hypertension. Garlic simmered 30 minutes in water and strain; this is mixed with coconut oil and is good for worm expulsion. The garlic tea, mixed with honey is good for whooping cough, fever and disorder of the nervous system. Garlic paste is externally applied over wounds and poultice for ringworms. Raw garlic juice is use for skin eruption and infection. Diluted in water the garlic juice is good for earache and prevention of grey hair. Raw garlic juice is good for whooping cough and tuberculosis.

4. Turmeric (Khasi – Shynrai)

Botanical name: *Curcuma domestica*

Medicinal uses

1. Relieve flatulence, gripping pain from stomach.
2. Heals wound and sprains.

3. Treatment of urinary diseases.

4. Treatment for rheumatism.

Methods - The rhizome is prepared as a paste or powder and mixed with milk, used as a drink to relieve flatulence or stomach pain. The paste is externally applied for sprain and wounds. The dried rhizome is simmering for 30 minutes in water and this water is used for liver and urinary disorder. The juice from the fresh rhizome is mixed with mustard oil and used for massage in rheumatism and pain of the nerves.

5. Spinach (Kh – Palong)

Botanical name: *Basella alba*

Medicinal uses

1. Soothes and protect the alimentary canal and relieve inflammation.

2. Promote the flow of urine.

3. Helps in softening inflamed part and lower the body heat.

4. Help in hypertension.

Methods - Consumption of cooked leaves has the property of soothing and protecting the alimentary canal, relieve inflammation, flow of urine and body heat. Leaves eaten raw help in hypertension.

6. Garden mint (Kh- Pudina)

Botanical name: *Mentha spicata*

Medicinal uses

1. Relieve flatulent, stomach pain. Used as a stimulant and tunic.

2. Treatment of vomiting.

3. Used for painful menstrual period.

4. Rid of intestinal worms.

5. Cooling effect and reduce itching of the skin.

Methods - Children should use mint tea to get rid of worms. Crushed leaves were used externally for cooling effect and itching of the skin.

7. Pennywort (Kh – Khlingsyiar)

Botanical name: *Centella asiatica*

Medicinal uses

1. Used in case of diarrhoea, dysentery and jaundice.
2. Purify blood and sooth the nerves.
3. Treatment of skin disease, sores, piles and tuberculosis.
4. Good for hair growth and check hair fall.

Methods - Whole plant possesses antibacterial and anti-inflammatory properties. The dried powder leaves help in tuberculosis. The plant is simmered in water and this is used externally to apply for falling of hair and promotes hair growth.

8. Black Pepper (Kh – Sohmrit)

Botanical name : *Piper nigrum*

Medicinal uses

1. Used for bladder complaint, kidney diseases, diabetes and dysentery.
2. Good for piles, tumor and boils. Promote flow of urine.
3. Treatment for chronic skin diseases.

Methods - Powdered pepper cure tooth-ache, it is also used as insecticide against clothes moths. Powdered leaves used for piles, tumor and boils. Powdered is mixed with oil and apply for chronic skin diseases. Pepper is used daily in food preparation for good flavour, but it also has a stomachic and anti-bacterial properties, it helps to cure vomiting, diarrhoea and colic resulting from cold.

9. Wood Apple (Kh - Soh Kwit)

Botanical name : *Feronia limonia*

Medicinal uses

1. Used in fever, inflamed mucous membrane.
2. Helps in asthma, jaundice, scurvy and dysentery.
3. Used for indigestion or pain in stomach.

Methods - The leaves are simmered in water, this is used as external application to control an inflamed mucous membrane. In case of asthma the juice from the crushed leaves is taken. The unripe fruit is eaten raw that helps the stomach in, indigestion, dysentery and prevent scurvy. The juice of ripe fruit is mixed with water and used in case of chronic constipation, indigestion, discomfort and pain. The juice of ripe fruit is mixed with oil and applied externally to control high fever and body inflammation as well in acute bronchitis

10. Lantana (Kh – Soh Pangkhlieh)

Botanical name: *Lantana camara*

Medicinal uses

1. Used as remedy for whooping cough, sprain, chronic inflammable skin.
2. Treatment of fever, jaundice.
3. Used as stimulant.

Methods - A paste with roots and leave using water is applied externally on the chest to relieve whooping cough or a chronic inflammation of the skin. The pounded fresh leaves are used for sprains. The flowering tops are soaked in water, this water is drunk in case of fever, cold, cough and jaundice.

11. Aloe

Botanical name: *Aloe vera*

Medicinal uses

1. Relieve muscle pain and soothes burn.
2. Heal wounds and cuts
3. Soften the swollen parts.
4. Kills or expel worms from the body.
5. It checks the growth of tumour, spleen problem, liver complaint, asthma and leprosy.

Methods - Juice from the fresh leaves is apply externally in case of burns, wounds, cuts and falling of hair. The leaf pulps are used in relieving muscle pain and kill worms.

12. Jack Fruit (Kh – Soh Phan)

Botanical name: *Artocarpus heterophyllus*

Medicinal uses

1. Relieve abdominal pain.
2. Curing of wound, swelling and insect bites.
3. Used for diarrhoea, fever and asthma.

Methods - Consumption of ripe fruits acts as a laxative. Consumption of seeds which are roasted or boil gives tone and vitality to the body and also relieve in abdominal pain. The roots are soaked in water; this water helps in diarrhoea, fever and asthma. From the tree, the milky juice collected is externally applied for curing of wounds, swelling and insect bites.

13. Borage

Botanical name: *Coleus aromaticus*

Medicinal uses

1. Used for sore, boil and sprain.
2. Used for sore throat and headache.

Methods - The juice of the leaves acts as a good ointment when applied externally for sores, boils, sprains and swelling.

The leaves are heated over hot water vapour and then applied externally to relieve sore throat. The fresh pounded leaves are applied externally to relieve head-ache.

14. Sensitive Plant (Kh – Bat Iambait)

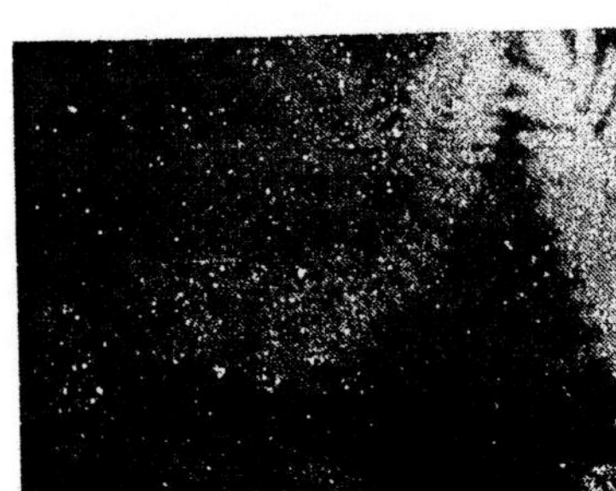

Botanical name: *Mimosa pudica*

Medicinal uses

1. Used for ulcers, skin diseases and swelling.
2. Used for mouth and throat problem.
3. Used in scorpion sting.
4. Used for treatment of mumps.

Methods - The juice of the plant is externally used as a lotion in ulcers and skin diseases. Leaves are crushed into paste and applied on the swelling or over the scorpion sting to reduce the pain and also used in mumps swelling. The seeds are crushed in water and this water is used to gargle in case of sore throat problem.

15. Banana (Kh – La Dew / Kait)

Botanical name: *Musa paradisiaca*

Medicinal uses

1. Stem juice is used in diarrhoea, dysentery, jaundice, wound and cuts.
2. Leaves used to heal wound and cuts.
3. Flowers check menstrual problem and earache.
4. Fruits used in diabetes, protect alimentary canal and promotes the flow of urine.
5. Relieves sore throat and chest congestion.

Methods - Juice from the plant obtained by squeezing used to drink raw help in dysentery, diarrhea and Jaundice or juice applies on cuts to stop bleeding. Young leaves used as a cool dressing for headache. In blister or inflamed surface, the leaves paste is applied to it. The juice of boil flowers is given in menstrual problem. The cooked flowers are eaten in case of diabetes. The sap of the flower is used for ear-ache. The unripe banana is good in diet in case of diabetes. The ripe banana cures or prevents scurvy, soothes and protects the alimentary canal, relieves inflammation, promotes flow of urine, chest congestion accompanied by cough. It is a special food in case of weakness and general loss of strength.

16. Sugarcane (Kh – Pai)

Botanical name : *Saccharum officinarum*

Medicinal uses

1. Used for jaundice, cold, fever and asthma.
2. Promotes flow of urine.
3. Help to strengthen the gums.

Methods - 3tsp of the juice is mixed with 2tsp honey, 1tsp ginger juice and ½ tsp of black pepper powder in a glass of water. This is to be taken twice a day; it is good for jaundice and asthma. The juice from the stem helps in the flow of urine and has a cooling effect. It also helps the gum.

17. Wild Ginger (Kh – Sying Khlaw)

Botanical name: *Zingiber zerumbet*

Medicinal uses

1. Treatment for fish poisoning.
2. Bacterial diseases and diabetes.
3. Cure stomach trouble and mouth infection.

Methods - The rhizome is made into a paste and mixed with water. This is used for cough, bacterial infection and diabetes. It also used in case of fish poisoning, to cure stomach pain and mouth infection.

18. Pomegranate

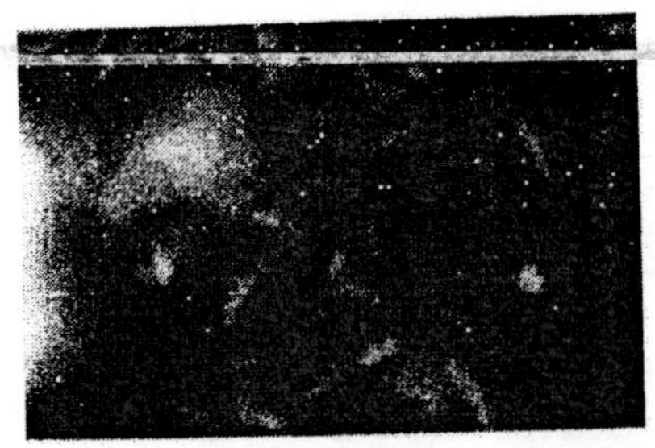

Botanical name: *Punica granatum*

Medicinal uses

1. Used to treat tapeworms and other worms in the body.
2. Used for curing dysentery and chronic diarrhoea.
3. Used to lower body heat and conjunctivitis.

Methods - Fresh leaves are collected and crushed, the juice obtained is used for dysentery. The peel of the fruit is chewed to cure dysentery and chronic diarrhea. Juice obtained by crushing the leaves and flowers is an ointment for conjunctivitis. The bark of the root and stem crushed together with water is used mostly for tapeworm expulsion from the body. This is to be taken three times a day. The seeds when grind into a paste is mixed with milk and is used for stone kidney. The fruit as a whole helps in strengthening of stomach and lowering body heat.

19. Gooseberry (Kh – Soh Mylleng)

Botanical name: *Phyllanthus emblica*

Medicinal uses

1. Used for tooth decay.
2. Help in flatulence and gastric.
3. Cure dysentery and diabetes.

Methods - The seeds are ground, then mixed with water, drink 1tsp in the morning and evening to help in the above uses.

Grind a fresh fruit and take only the juice for dysentery remedy.

20. Kh - Kynbat Ïong

Botanical name : *Eupatorium adenophorum*

Medicinal uses

1. Used for cuts and wounds.

Method - Crush the leaves and mix it with lime and apply externally on wounds or cuts.

MIXTURES

Mixture – I: Kynbat Lynnai, Kynbat Tyrphin, Kynbat nit

Treatment - This Mixture is widely used locally by the newborn baby and mother instead of Bonisan or Gripe water.

Mixture – II: Kynbat Synsar, Kynbat Tyrphin, Kynbat Rit, Kynbat Ïong,

Kynbat Lai pnat, Kynbat bseiñ, Kynbat Jingbuiñ Khla, Bniat Hati.

Treatment - This Mixture is used for any poison treatment. Also used in case of dysentery.

Mixture –III : Kynbat Tohjaw, Sohmon

Byrtan, Peach Leaves.

Treatment - For any maggot-infected part of human body, wounds and cuts.

Table 1
Indigenous remedies for treatment of digestive disorders

Disease	Name of the plant	Botanical Name	Properties	Scientific basis	Method of use/ preparation
Diarrhoea	Coriander	*Coriandum sativum*	Stomachic Constipating Carminative Refrigerant Tonic	Improve digestion Control diarrhoea Cures flatulence Provide soothing effects Invigorates body tone	Soak 2 or 3tsp coriander seeds overnight & take next morning with 1-cup of buttermilk.
	Ginger	*Zingiber officinale*	Carminative Anodyne Anthelmintic Appetizer Tonic Emollient	Cures flatulence Allays pain Destroys germs Stimulates appetite Provide strength Provides soothing effects	Take equal amounts of ginger & black pepper and then make a fine powder.
	Isabgol	*Plantago ovato forsk*	Robarant Stomachic Constipating	Provide strength Relieves pain in stomach Control diarrhea	Mix 2tsp seeds of isabgo l in ½ cup of curd.
Dysentery	Coriander	*Coriandum sativum*	Stomachic Stimulant Anthelmintic Carminative Antispasmodic	Increase appetite Stimulates body activity Extricate worms from body Cures flatulence Release from spasm	Grind fresh leaves to extract juice, mix 1-2tsp fresh juice in 1cup of buttermilk.
	Isabgol	*Plantago ovato forsk*	Emollient Demulcient Laxative	Provide soothing effect Provide calm effect Loosen the bowels	Take 2tsp of seed husk with 1 glass of milk.

Disease	Name of the plant	Botanical Name	Properties	Scientific basis	Method of use/ preparation
Constip-tion	Amla	*Phyllanthus emblica*	Laxative Digestive Stomachic Carminative Relieve pain in stomach Cures flatulence	Loosen the bowels Improves digestion	Take 50gram of dried fruit and grind into a fine paste.
	Date	*Phoenix dactylifera*	Laxative Digestive Tonic	Loosen the bowels Improves digestion Acts as a tonic	Soak 100gm. Dates in water for overnight, boil for a few minutes.
Stomach pain	Amla	*Phyllanthus emblica*	Anodyne Digestive Stomachic	Allays pain Improves digestion Relieves pain in stomach	Soak dried amla in 1glass of water for overnight, strain the water in the morning and drink it.
Vomiting	Mint	*Mentha arvensis*	Anti emetic Antiseptic Refrigerant Anthelmintic	Controls vomiting Extricates worms Provide cooling effect Destroys germs	Extract juice from the leaves and drink.
	Ginger	*Zinggiber officinale*	Anti emetic	Controls vomiting	Mix well ½ tsp of fresh ginger juice, 1tsp of lime juice, 1tsp mint juice and 1tsp of honey

Conclusion

The rural communities especially the remote areas rely greatly on indigenous knowledge for health care. Since ages herbs are being used as medicine as they are readily available, safe and cost effective. In spite of the overwhelming influence and over-dependence on modern medicine, many people still rely on herbal drugs. The reason is that, if the herbal medicines are used properly they don't have any side effect. Thus effort should be made to promote such traditional knowledge of healing for better health.

Plants still constitute one of the major raw materials for drugs for treatment of various ailments of human being, although there has been significant development in the field of synthetic drugs, chemistry and antibiotic. During the last decades considerable changes have taken place in the medicinal system all over the world, because of the general awareness of the widespread of toxicity and harmful after effects associated with the long use of synthetic drugs and antibiotics.

REFERENCES

Joseph J. and Kharkongor P. 1997. A preliminary ethnobotanical survey in Khasi and Jaintia Hills, Meghalaya: *Contribution to Indian Ethnobotany* (ed. S.K. Jain) 187 – 194.

Kharbuli, B,. Kayang, H and Syiem, D. 2006. *Biodiversity in Northeast India.* NEHU, Publication.

Kharduit J. 1999. Medicinal Plants: Utilization and conservation in Meghalaya. *Biodiversity Northeast India Perspective* (ed. B.Kharbuli, D. Syiem and H. Kayang): 89 – 92.

16

Sustainable Utilization of Medicinal Plants with Special Reference to Psoriasis by Using an Indigenous Plant Found in Northeast India

Sumita Kharmawphlang

1. Introduction

Medicinal plants are the local heritage with global importance. The world is endowed with a rich wealth of medicinal plants. Herbs have always been the principal form of medicine in India and presently they are becoming popular throughout the developed world, as people strive to stay healthy in the face of chronic stress and pollution, and to treat illness with medicines that work in concert with the body's own defenses. Medicinal plants also play an important role in the lives of rural people, particularly in areas where modern health facilities are not available.

Medicinal plants have curative properties due to the presence of various complex chemical substances of different compositions, which are found as secondary plant metabolites in one or more parts of these plants. These plant metabolites, according to their composition are grouped as alkaloids, glycosides, corticosteroids, essential oils, etc. The alkaloids form the largest group which include morphine and codein (poppy), strychnine and brucine (*Nux vomica*), quinine (*Cinchona*), ergotamine (ergot), hyocyamine (bellodona), scolapomine *(Datura)*, emetine (ipemac), cocaine (coco), ephedrine (*Ephedra*), reserpin (*Rauvolfia*), Caffeine (tea-dust), aconite, vascine (vasaca), santonin (*Artemisia*), lobelin (lobolia) and many others. Glycosides form another important group represented by digoxin (foxglove), stropanthin (*Strophanthus*), glycyrrhizin (liquorice), barbdin (*Aloe*), sannocides (senna), etc. Corticosteroids are included in sennocides (senna), etc. and have come into prominence recently. Diosgenin (*Dioscorea*) and solasdin (*Solanum* sp.) etc. now command a large world demand.

Some essential oils such as those of valerian, kutch and peppermint also possess medicating properties and are used in the pharmaceutical industry. However, it should be stated in all fairness that our knowledge of the genetic and physiological makeup of most of the medicinal plants is poor and we still know very less about the biosynthetic pathways leading to the formation of active constituents for which these crops are valued.

India enjoys the privilege of having time tested traditional systems of medicines based on the natural products. Plant based products have been used for medicinal, therapeutic or other purposes right from the dawn of history. The traditional Indian System of Medicine, namely Ayurveda which involves dispensing of herbal and plant products in various forms such as powders, extracts, decoctions etc., dates back to the Vedic period, when the first mention of diseases and drugs are found in the *Rigveda* and *Yajurveda* (i.e. around 2000BC) and the earliest comprehensive description of Ayurvedic is available in Atharvaveda (i.e. around 1600-1000BC), which contains inter alia descriptions of human anatomy, rudiments of classification of diseases and reference to herbal medicine. With the revival of Indian traditional medicinal systems like Ayurveda, the domestic demand for medicinal plants has seen a rapid increase and the economic importance of these plants has increased enormously. Conservation of medicinal plants is virtually important for the maintenance of biodiversity and the preservation of indigenous knowledge.

India possesses almost 8% of the estimated biodiversity of the world around 0.126 million species. India is one of the mega biodiversity centres of the world in which the important regions are the Western Ghats and the Northeastern Region.

In India, there are 7,843 licensed pharmacies of Indian System of Medicines and 857 of Homeopathy (ISM and H) and a number of unlicensed small scale processing units engaged in the requirement of 4.65 lakhs registered practitioners of ISM and H and other users. The folklore practitioners are included as the other users too.

About 8,000 estimated species of plants are used in human and animal health care and over 10,000 herbal drug formulations have been recorded in codified medical texts of Ayurveda. Some of the important medicinal plants mentioned in Ayurvedic texts which are available in the Northeast region particularly in Meghalaya are *Allium saturin, Acorus calamus, Aconitum heterophyllum, Aconitum ferox, Asparagus racemosa, Azadirachta indica, Berberis aristata, Rubia cordifolia, Solanum khasianum, Swertia chirata, Taxum wallichiara, Tinospora cordifilia, Withania somnifora* etc.

2. Methods and Results

1. *Allium Sativum*

Botanical name	:	*Allium sativum*
Family name	:	Liliaceae
Sanskrit name	:	Rasona
Common name	:	Garlic

Khasi name : Rynsun
Part used : Bulb

Chemical composition

The active principle named allin and a number of other antibiotic principles have been isolated, namely, allistatin I, allistatin II and garlicin.

Medicinal Value of Garlic

Ayurveda claims a lot of therapeutic value of garlic or Rasona and it considers as an important plant. To give a scientific basis for the ancient claims regarding the therapeutic efficacy of garlic in various diseases and disorders, a large number of experiments have been carried out and as a result garlic has been found to be of value in conditions like-

- Diabetes
- Hypertension
- Arteriosclerosis
- Angina pectoris
- Chronic colitis
- Gastritis
- Rheumatoid arthritis
- Helminthiasis
- Amoebiasis
- Bacterial and Fungal infections.

Table 1

According to Ayurveda, the medicinal uses of Garlic are given below:-

Sl. No.	Disorder of	Uses	Based on Ayurvedic theory
1	Gastro-intestinal tract	Carminative, antiflatulent. appetizer, digestive	Because of Ushna Veerya/ potency
2	Respiratory tract	Expectorant	Because of Ushna veerya, it balances the aggravated kapha dosha
3	Miscellaneous type	Anti-septic, anti-rheumatic, diuretic	Because of ushna veerya, it acts as anti-inflammatory, since aggravated vata causes pain.

Methods of using as home remedies

- Single clove on empty stomach for hypertension, arteriosclerosis.
- Add few cloves during cooking since it helps in digestion.
- Warm a few cloves and mustard oil and apply locally on Rheumatoid arthritis.

2. Aconitum Ferox

Botanical name	:	*Aconitum ferox* Wall
Family name	:	Ranunculaceae
Common name	:	Bikh
Parts use	:	Roots

Medicinal Value

It is alterative, diaphoretic, diuretic, expectorant and febrifuge. It is useful in diabetes, debility, asthma, ear and nose discharges, leprosy and paralysis. Its powdered form is administered orally in rheumatism and fever.

3. *Aconitum Heterophyllum*

Botanical name	:	*Aconitum heterophyllum*
Family name	:	Ranunculaceae
Sanskrit name	:	Ativish
Common name	:	Atis
Parts used	:	Roots

Chemical composition

The roots are medically important and contain amorphous alkaloids, atisine, dehydroatisine, heteratisine, hetisine, aconitic acid, tannic acid, pectin and starch.

Medicinal Value

- It is used as antiarthritic agent, astringent, carminative, febrifuge, stomachic and aphrodisiac.
- Powder of the root with honey is given to the child in small doses in case of fever, diarrhea, indigestion, vomiting and cough.

It is important to note that for internal use, roots of ativisha should be purified.

4. *Asparagus Rachemosa*

Botanical name	:	*Asparagus racemosa*
Family name	:	Asparagaceae
Sanskrit name	:	Shaṭavari
Common name	:	Shatavar
Parts use	:	Roots

Medicinal value

Uses in dysentery, as diuretic and tonic especially in post partum and lactating mothers

5. *Azadirachta Indica*

Botanical name	:	*Azadirachta*
Family name	:	Meliaceae
Sanskrit name	:	Nimba
Common name	:	Neem
Parts use	:	Whole plant

Chemical composition

Neem flowers consist of 3 related compounds- kaempferol, quarcetin and myricetin. The trunk bark contained nimbin, nimbinin, nimbidin, nimbesterol, essential oil, tannin and a bitter principle, margosine. Neem oil contains various acids, sulphur etc. Preliminary chemical examination of the oil indicated the presence of sulphur. The principal constituents of the oil were sulphur, alkaloid, resins, glycosides and fatty acids. Hot aqueous extract of crushed neem seeds on removal of the albuminous and other foreign matter by basic lead acetate yields white non-bitter sulphur. The flowers yield nimbosterol. Nimbosterol could be obtained from other parts as well.

Neem oil and two of its bitter principles nimbidin and nimbidol were studied for antibacterial effects. The seeds of the plant were found to contain 30% of oil that is bitter in taste. A series of compounds were isolated e.g. nimbidinic acid, of which the sodium salt was found to be an active compound. Sodium nimbidinate showed diuretic effect. Nimbidin was found to be effective in various skin diseases such as furunculosis, arsenial dermatitis, ulcers due to burns, herpes labiasis infection, scabies and seborrhoeic dermatitis. Nimbidin gargle and dentrifrices have been found to be effective in the treatment of bleeding gums.

Medicinal value

Used in skin diseases, intestinal worms, piles and urinary diseases. Root- bark and young fruits are astringent, tonic and antiperiodic. Bark is bitter, tonic, astringent, antiperiodic and vermifuge. Fruit is purgative, emollient and antihelminth. Leaf juice is antihelminth. The oil extracted from nuts and leaves acts as a local stimulant, antiseptic and insecticide.

Dose:

Bark powder	:	2-4 gms
Leaf juice	:	10-20 ml
Patent drugs	:	Nimbadi churna, Nimbadi arishta and imbaharidra khanda

6. *Rubia Cordifolia*

Botanical name	:	*Rubia cordifolia*
Family name	:	Rubiaceae
Sanskrit name	:	Manjishta
Common name	:	Indian madder
Parts use	:	Roots

Chemical composition

Its components are purpurin and manjishtin.

Medicinal value

It is used in discolouration of the skin and diabetes. It reduces inflammation and acts as antipyretic, blood purifier and antihelminth.

Dose:

Decoction (Kwatha)	:	50-100ml
Powder	:	1-3gms.
Caution	:	Increase in dosage causes sedative action.
Patent drugs	:	Manjishta kwatha, Mabjishtvardhaka

7. *Swertia Chirata*

Botanical name	:	*Swertia chirata*
Family name	:	Gentiaceae

Sanskrit name	:	Kiratatikta
Common name	:	Chirayata
Parts use	:	Whole plant

Medicinal value

Used for washing of wounds, in liver disorders, jaundice, respiratory disorders, skin disorders, fever, as anti-pyretic and antihelminth.

Dose:

Decoction	:	50-100 ml
Powder	:	1-3 Gms
Patent drug	:	Sudarshan churna, Kiratadi kwatha

8. *Tinospora Cordifolia*

Botanical name	:	*Tinospora cordifolia*
Family name	:	Menispermaceae
Sanskrit name	:	Guduchi Amritavalli
Common name	:	Guduchi
Parts use	:	Stem

Chemical composition

Stem contains tinosporin and a furanoid diterpene, giloin, gilonin, berberine and a waxy substrate.

Medicinal value

It is used as alterative, antihelminth, anti-arthritic, antiperiodic, antipyretic, aphrodisiac, bitter tonic, blood purifier, cardiac, carminative, digestive, diuretic and expectorant. It is used in gouty arthritis, rheumatoid arthritis, jaundice, diabetes, skin diseases and fever. In fever during pregnancy, a decoction of the stem is taken internally twice daily on empty stomach.

9. *Withania Somnifera*

Botanical name	:	*Withania somnifera*
Family name	:	Solanaceae
Sanskrit name	:	Ashwagandha
Parts use	:	Roots

Chemical composition

Roots contain an essential oil ipuranol, a crystalline alcohol, whithaniol, withaninine, somniferene, hentriacontane and phytosterol.

Medicinal value

It is used as alterative, aphrodisiac, tonic, diuretic, narcotic, hypnotic, sedative, restorative, abortificient, rheumatism, cough and general weakness. It is also indicated in hypotensive and depressant, anti-spasmodic, asthma, anti-tumour action. The powder of dried root is given with milk or boiled with milk as a general tonic especially during pregnancy or after delivery if the woman is underweight or lethargy and also to lactating mothers to enhance milk secretion. In the similar way of preparation it is given to children.

Dose:

Powder	:	- 3-6gms
Patent drug	:	Ashwagandha churna, Ashwagandharasayana, Ashwagandha ghrita and Ashwagandharishta

All the plants that have been mentioned above are important and constitute some of the important raw materials in pharmaceutical companies. These plants need conservation since some of them are endangered species. Some of the few common plants will be worth mentioning for the benefit of both the rural and urban population. As a matter of fact, we all could play a role in conserving medicinal plants either by planting in our backyards, kitchen garden or as potted plants. These plants can be used as home remedies for simple ailments that we come across frequently, which we could deal it by not allowing ailments to aggravate.

Some of the common, yet important medicinal plants are as follows:-

10. *Aloe Vera*

Botanical name	:	*Aloe vera*
Family name	:	Liliaceae
Sanskrit name	:	Kumari, Ghritakumari
Common name	:	Indian Aloe
Parts use	:	Fleshy pulp

Common uses

- The fleshy pulp helps in healing of wounds. The pulp is tied over the wound in daily dressing, since it hastens wound healing.

- In burns, apply the fleshy pulp to the area.
- Cosmetic purposes: peel out the skin and take the pulp, apply over the face as a facemask, leave for 10-15 minutes. Wash with cold water.

Patent drug	:	Kumaryasava

11. *Apium Graveolens*

Botanical name	:	*Apium graveolens*
Family name	:	Umbelliferae
Sanskrit name	:	Ajamoda
Common name	:	Celery
Parts use	:	Whole plant (in Ayurveda, seeds are used).

Celery is an important salad plant that has been cultivated from more than 2000 years. Since it is considered as a salad crop, it is mostly grown in kitchen or home gardens. Celery is an excellent basic food. It is one of the best sources of minerals and vitamins. An analysis of celery leaves consists of moisture, proteins, minerals, fibres and carbohydrates. Its mineral and vitamin content are calcium, phosphorus, iron, carotene, riboflavin, niacine and vitamin C. The fruits yield a pale yellow volatile oil. It consists of limonene d-selinene sedanonic acid, anhydride and sedanolide.

Common uses

- Arthritis:- Rheumatoid, osteoarthritis, gouty arthritis
- Juice of celery in combination with carrot juice helps in anemia, leukemia, haemophilia. Useful in asthma, bronchitis and tuberculosis
- The seeds of the celery are effective in indigestion. A teaspoonful of the seeds are soaked in a glass of butter milk for 5-6 hours, should be grounded in the same butter milk mixture and administered to relieve indigestion.
- Celery is a valuable food for those who are prone to stone formation in the Gall bladder or kidneys; its regular use prevents stone formation.
- The powder of the dried root mixed with a tablespoon of honey is taken twice daily as general tonic.
- Celery juices mixed with a tablespoon of honey, taken at night before retiring helps in relax and soothing sleep.

Since Celery has a well-balanced content of the basic minerals, vitamins and nutrients, inclusion in our diet as flavors to stews and salads will provide our body with essential nutrients.

12. *Ocimum Sanctum*

Botanical name	:	*Ocimum sanctum*
Family name	:	Lamiaceae
Sanskrit name	:	Tulsi, Krishna tulsi
Common name	:	Holy basil
Parts use	:	Leaves, roots, seeds

The leaves and seeds yield an essential oil containing eugenol, carvacrol, methyl eugenol and caryophyllene. The oil possesses antibacterial and insecticidal properties. Seeds are mucilaginous, demulcent, and useful for urinary system. The juice or infusion of leaves possesses diaphroretic, antiperiodic, stimulating and expectorant properties. It is used in catarrh and brochitis, applied to skin in ringworm and other cutaneous diseases.

Common uses

- The juice of leaves mix with honey is good for cough.
- The juice of leaves is also indicated in fever

Dose:

Juice of leaves	:	10-20ml
Powder	:	3-6gms

13. *Adhatoda Vasica*

Botanical name	:	*Adhatoda vasica*
Family name	:	Acanthaceae
Sanskrit name	:	Vasa, Vasaka
Common name	:	Malabar nut
Parts use	:	Leaves, Roots, flowers

Adhatoda vasica is a well-known medicinal plant found commonly in all parts of India. Leaves and flowers contain alkaloids vasicine, vasicinine, alpha-sitosterol. Kaempferol and vasicinone has been isolated from the plant.

Common uses

Leaves and roots are useful in cough, asthma, bronchitis, rheumatism and as insecticidal. Flowers, leaves and roots are antiseptic, antispasmodic, alleviative, blood purifier and tonic. For immediate expectoration in children and spasmodic attack of asthma, fresh juice gives immediate relief which may be followed by other preparations of the drug. The twigs of the plant when use as tooth picks, offer relief for pyorrhea. For rheumatoid arthritis, a paste of leaves is applied locally.

Dose:

Leaf juice	:	10-20 ml
Patent drug	:	Vasavaleha, Vasarishta and Vasachandanadi taila.

14. *Vitex Negundo*

Botanical name	:	*Vitex negundo*
Family name	:	Verbenaceae
Sanskrit name	:	Nirgundi
Parts use	:	Leaves, roots, seeds

Chemical composition

The fresh leaves yield an essential oil and the bark yields flavones glycosides. The leaves possess antiarthritic, antiparasitic and aromatic properties.

Common uses

It is useful in curative affects in asthma, lung diseases, spleen enlargement, urinary disorders, tonsillitis, sciatica and rheumatism.

- A pillow stuffed with the leaves of this plant is placed under the head for relief of headache.
- For rheumatoid arthritis especially with join swelling, the heated leaves are applied over the area and a piece of cloth is tied around it and kept for a while, it can be repeated a number of times till the person is relief of pain.

Dose:

Leaf juice	:	10-20ml
Patent drug	:	Nirgundi kalpa, Nirgundi taila.

Psoriasis is a less common, but chronic, pathetic and non-infectious skin disease characterized by well-defined slightly raised, dry erythematous macules with silvery scales and itching. Psoriasis is found in different parts of the world varying from 0.1 to 3.0%.

Psoriasis is also considered to be heredo-familial disease. According to Ayurvedic classics, kushta have been classified under Adibala Pravritta Vyadhis. Charaka also mentioned that dushti beeja is also considered as the aetiology of kushta disorder. Psoriasis is a genetically determined disorder. But these factors may remain dormant and flare up when there is a triggering factor.

The method selected for psoriasis treatment was conducted in the present study using both systemic or internal medicines and topical or external medicaments that have least or no side effects. The internal medicines selected are Arogyavardhini vati, Panchatiktaghrita guggulu and Amypure syrup that are indicated for kushta (skin disease). The medicine used for external application is the trial drug extract (extract in coconut oil) of *Wrightia tomentosa* leaves, which is found abundantly in Northeast India.

The present investigation reveals that the trial drug has a significant role in controlling psoriasis. The overall response of the trial group after two months of treatment showed that 66.67 percent of the total cases (25 cases) obtained complete relief, 27 percent with marked relief and only 7 percent with moderate relief. The results of this findings indicated that the antimicrobial activity of *Wrightia tomentosa* might be attributed to the presence of tryptanthrin, antileukemic activity of indirubin and the antioxidant activity of rutin, indigotin and indirubin. The antioxidative, antiproliferative and antimicrobial principles of *Wrightia tomentosa* would seem to have accounted for its therapeutic effect in the treatment of psoriasis.

A comparative study between the trial group and control group was also studied in order to have a better picture on the efficacy of the trial drug in the treatment of psoriasis cases. The statistical test of significance was carried out between the two groups and showed that there is a significant difference in response to the treatment. During the entire course of treatment, it was noted that in the trial group all the patients treated with the trial drug did not have a relapse even after a follow up of one month. But a follow up study was continued for another four months to confirm its efficacy and after which no relapses were seen in all the patients. This confirms the sustained and excellent therapeutic efficacy of this simple drug that might also be cost effective.

3. Conclusion

- Conservation of endangered species of medicinal plants found in NE region of India is the need of the hour.
- With the increasing demand of raw materials in the country and abroad, cultivation of important medicinal plants in the region is essential.

- Opening up of pharmaceutical companies in the state.
- Conservation of common medicinal plants by planting in the backyard or as potted plants, for using as home remedies in simple ailments.
- Potency of the medicinal plants found in Northeast region should be evaluated as those found in other parts of India.
- Indepth, analytical and clinical study of indigenous plants found in Northeast region should be encouraged.
- Meghalaya is fortunate that there are still many people with rich knowledge of folklore medicine and traditional healing methods, their work should be encouraged and methods for analyzing and standardizing these medicines can be developed.

REFEREENCES

Sharma P. V. 1994. *Dravya Guna Vignana*, Vol II. Chaukhambha Bharati Academy, Varanasi.

Kurian J.C. 1999. Plants that Heal. Oriental Watchman Publishing House. Pune.

Das Narayan Prajapati and Purolit S. S. 2003. *Agro's Colour Atlas of Medicinal Plants*. Agrobios.

Panda H. *Medicinal Plants Cultivation and their Uses*. Asia Pacific Business Press Inc.

Medicinal Plants of India: Vol I. 1976. Indian Council of Medical.Research. New Delhi.

Medicinal Plants of India: Vol I, ICMR. New Delhi.

Zafar Rasheeduz. 2004. *Medicinal Plants of India*. CBS Publishers and Distributors.

Venugopal S. N. 1992. *Medicinal Plants Useful in Mother and Child Health*. Directorate of Advertising and Visual Publicity, Ministry of Iand B, for the Department of Family Welfare, Ministry of Health & Family Welfare, Government of India, New Delhi.

CUTS . 2004. *Database on Medicinal Plants*. 1-2; 15. R.K.Printers, Calcutta.

Kharbuli B and Marngar D. *Conservation of Medicinal Plants Diversity of Meghalaya*.

17

Taxonomy and Industrial Potentiality of *Cinnamomum parthenoxylon* Meissn. Growing in Meghalaya, Northeast India

Akhil Baruah

1. Introduction

Cinnamomum parthenoxylon Meissn., belonging to the family Lauraceae, is a moderate to large sized evergreen and aromatic tree, distributed in tropical to sub-tropical regions of Burma, China and India (Hooker 1885, Kurz 1877, Zhu *et al.* 1994). Almost all parts of the plant yield seven different types of essential oils viz. (1) 1, 8-cineole, (2) D-camphor, (3) farnesene and nerolidol/9-oxonerolidol, (4) D-linalool, (5) citral, (6) monoterpine, a-pinene and terpinen, and (7) methyl eugenol (Zhu *et al.* 1994).

In India, *C. parthenoxylon* is occurring mainly in the Northeastern region up to an altitude of 1,750m (Kanjilal *et al.* 1940). In an ethno-floristic survey followed by the chemical examinations of plant parts of the *Cinnamomum* members available in Northeast India during 1994-1997, two plant populations of the species (*C. parthenoxylon*) with different odoriferous characters were detected from Meghalaya state. The plant parts of one population (Variant I) possess odor similar to bottle brass (*Callistemon lanceolatus* DC.) leaf oil, while the same plant parts of other population (Variant II) possess odor similar to linalool type-sweet basil (*Ocimum basilicum* L.) oil (Baruah 2000, Baruah & Nath 2000). In the present communication, the morphology including significant foliar micro-morphology of epidermis and venation, phenology, distribution and ecology, local name and uses, leaf and stem bark essential oils followed by a conclusion with regards to the taxonomic status and industrial potentiality of *C. parthenoxylon* variants growing in

Northeastern part of India is reviewed (Baruah 2000, Baruah & Nath 2000, Baruah *et al.* 2000).

2. Methods and Results

Morphological Characters (Variant I)

A large sized evergreen tree with open conical crown; Bark smell unlike that of the Variant II, dark grey, deeply ridged and fissured; Twigs slender, glabrous; Buds small, round with many silky scales; Leaves smell like stem bark, alternate, coriaceous, glabrous, glossy above, dark green, pale below, ovate to broadly elliptic or elliptic-oblong, apex acuminate to sub-caudate, usually rounded-cuneat at base, variable in size, 3-6.5 x 6.5-14.5cm, base although triplinerved, exhibits both arodromous and brochydodromous nature, basal nerves often with puncted glands at the axils, lateral nerves (2° veins) on either side of the midrib 4-6; Petiole slender, 1-2.2cm long; Panicles sub-terminal to solitary axillary, nearly glabrous, glabrate with age, shorter than leaves, up-to 8.5cm long; Flowers pale-yellow, 4.5-5mm long, perianth 3+3, sub-equal, 1.75-2mm long, elliptic, minutely pubescent on both surfaces; Stamens 3+3+3, 1-1.5mm long, anther 4-locular, introrse, whorl III extrorse, valve open at dehiscence, filaments silky puberolous, glands of whorl III reniform, shortly stalked, yellowish, attached 1/3 of the base of the filament; Staminode 3, 0.75mm long, head hastate with bluntly acutish apex; Pistil 2mm long, ovary ellipsoid, minutely pubescent; Fruits globose, the size of a small pea, succulent, dark purple when ripe, seated on the expanded cup-shaped perianth.

Phenology: Flowerings – March to May; Fruiting - June to September.

Distribution and Ecology: Found sporadically in the evergreen forests, in and around Shillong of Meghalaya between the altitudes 1150-1450m.

Local Name: Kapur.

Use(s): Leaves and stem bark are aromatic with an agreeable odour and used for burning as incense.

Morphological Characters (Variant II)

A large sized evergreen tree similar to that of the Variant I, in respect of its morphological and qualitative foliar micro-morphological characters including the phenology, ecology and distribution, local name and use. However, the aromatic leaves and stem bark possess odour similar to that of "linalool type-sweet basil".

Significant Foliar Epidermal and Venation Characters

The quantitative data of foliar epidermal characters of the variants are presented in Table 1. No remarkable variations on foliar epidermal quantitative data are recorded between the variants. Epidermal cells tetragonal to polygonal and straight on upper surface while almost straight on lower surface. Leaves are hypostomatic and stomata are paracytic. In Variant I, number of stomata/mm^2 and stomatal index are 342 and 16.86 respectively, while in Variant II, number of stomata/mm^2 and stomatal index are 334 and 16.82 respectively.

The quantitative data of foliar venation characters of the variants are presented in Table 2. Except the size of areoles, their average frequency and absolute areoles number, no significant differences on venation quantitative data are revealed by the variants. The major venation pattern in both the variants is of acrodromous-brochidodromous in nature. The areoles are tetragonal to polygonal in shape. The size and number of areoles are variable at apical, middle and basal regions of the lamina. The vein and veinlet endings are simple and areoles are tetragonal to polygonal. In Variant I, average frequency of areole/mm^2 is 18.40, while in Variant II it is recorded as 10.67.

Essential Oils Characters of Leaf and Stem Bark

Almost all parts of the plant are aromatic and yield volatile oil. Physico-chemical characters like yield, colour, odour and refractive index of the oils of the variants are shown in Table 3. The leaf and stem bark essential oil components identified for the variants are presented in Table 4.

On GC analysis of the oils of Variant I, 12 components representing 99.91% of the total oil of the leaves are reported, where 1, 8-cineole alone constitutes 64.70%. The other components of above 3% concentration in the leaf oil are a-pinene (3.26%), b-pinene (14.83%) and a-terpineol (11.68%). Likewise, 20 components representing 96.77% of the total oil of the stem bark are identified where 1, 8-cineole alone constitutes 53.00%. The other components of above 3% concentration in the bark oil are camphor (3.08%), terpinen-4-ol (16.40%), a-terpinolene (16.40%) and d-elemene (3.03%).

On GC analysis of the oils of Variant II, 13 components representing 94.83% of the total oil of the leaves are reported, where linalool alone constitutes 86.9%. The other components of above 1% concentration in the leaf oil are terpinolene (1.81%) and a-humulene (2.85%). Likewise, 19 components representing 96.83% of the total oil of the stem bark are identified where linalool alone constitutes 87.12%. The other components of above 1% concentration in the bark oil are 1,8-cineole (1.35%), b-caryophyllene (1.40%) and caryophyllene oxide (1.15%).

Considering the quantitative occurrences of the individual oil components, there exist significant differences between the leaf and stem bark oils of the two variants. Variant I contain remarkable higher amount of 1, 8-cineole and lower amount of linalool in their oils in comparison to the Variant II. Significant difference was also observed on the quantitative occurrence of b-pinene in the oils of the variants. Likewise, a-terpinolene detected in the oils of the Variant I was at 11.68 - 16.40% concentration. However, the same component could not be detected in the oils of the Variant II.

Table 1
Foliar epidermal quantitative data of *Cinnamomum parthenoxylon* variants/ chemotypes

Variant	Epidermal cell Length (μm)		Epidermal cell Breath ((μm))		Epidermal cell/mm		[2]Stomata/ mm^2		Stomatal index	
	US	LS	US	LS	US	LS	US	LS	US	LS
Variant – I	35.91	33.25	22.57	17.35	1324	1686	-	342	-	16.86
Variant-II	36.12	32.97	21.92	18.07	1364	1654	-	334	-	16.82

US=Upper Surface, LS=Lower Surface, "-"=Absent

Table 2
Foliar venation quantitative data of *Cinnamomum parthenoxylon* variants/ chemotypes

Variants	Leaf area (mm^2)	No. of 2^0 veins on one side	Angle between 1^0 & 2^0 veins	Veins islets (areoles/ mm^2)	Veinlet entering in areoles/ mm^2	Veinlet termination number/ mm^2	Size of areoles (mm)	Absolute areoles number (in thousand)	Absolute vein entering number (in thousand)
Variant I	2967	4-6	30^0-60^0	A 20.0	1.2	1.2	0.26X0.13	54.60	4.70
				M 17.0	1.8	1.8	0.33X0.23		
				B 18.2	1.8	1.8	0.30X0.19		
Variant II	3744	4-6	30^0-60^0	A 11.0	1.6	1.6	0.32X0.22	39.94	6.49
				M 10.2	1.8	1.8	0.37X0.24		
				B 10.8	1.8	1.8	0.36X0.23		

A=Apical region, M=Middle region, B=Basal region.

Table 3

Physico-chemical characters of oils obtained from *Cinnamomum parthenoxylon* variants/chemotypes

Characters	Leaf oils		Stem bark oils	
	Variant I	Variant II	Variant I	Variant II
Oil yield (FWB)	3%	5%	0.2%	0.5%
Colour	Colourless	Colourless	Pale yellow	Colourless
Odour	Pleasent	Sweet	Pleasent	Sweet
Refractive index (25⁰C)	1.4622	1.4716	1.4703	1.4712

Table 4

Percentage composition of the essential oils of *Cinnamomum parthenoxylon* variants/chemotypes

Components	Variant I		Variant II	
	Leaf	Stem bark	Leaf	Stem bark
Benzaldehyde	0.45	0.50	—	—
a-pinene	3.26	1.35	0.24	0.56
Camphene	—	t	0.14	0.35
b-pinene	14.83	1.15	0.16	0.40
Myrcene	0.80	1.00	0.06	0.10
a-phellandrene	—	—	—	0.06
p-cymene	—	0.90	0.54	—
1,8-cineole	64.70	53.00	0.20	1.35
b-phellandrene	0.34	1.60	—	0.44
g-terpinene	0.90	—	0.50	—
Terpinolene	—	0.70	1.81	—
Linalool	0.70	t	86.90	87.12
Camphor	—	3.08	—	—
Borneol	0.90	1.44	—	0.50
Terpinen-4-ol	0.65	6.06	—	0.72

(Contd...)

Components	**Variant I**		**Variant II**	
	Leaf	**Stem bark**	**Leaf**	**Stem bark**
a-terpinolene	11.68	16.40	—	—
d-elemene	—	3.03	—	—
Llinalylacetate	—	—	—	0.44
Cinnamaldehyde	—	—	—	0.21
Eugenol	—	—	—	0.57
Methyl cinnamate	—	—	—	0.70
b-caryophyllene	—	0.60	0.30	1.40
Ethyl cinnamate	—	—	—	0.09
Methyl isoeugenol	—	—	0.20	0.50
a-humulene	0.80	1.45	2.85	—
Bycyclogermacrene	—	1.60	—	—
a-farnesene	—	1.90	—	0.17
Caryophyllene oxide	—	0.65	0.93	1.15
Total t=<0.05%	99.90	96.77	94.63	96.83

(Source: Baruah & Nath, 2000)

3. Conclusion

The investigation of Baruah & Nath (2000) authenticated the odoriferous differences between the two variants of *C. parthenoxylon* indicating the variants as the source of two different essential oils. One (Variant I) is rich in 1, 8-cineole, while the other (Variant II) in linalool. The finding thus indicates the availability of two promising sources of essential oil from Northeast India for possible commercial use.

Besides the odoriferous differences, the two variants also differ in their quantitative foliar venation characters like average frequency of areoles, their size and absolute areoles number. Except these above differences, the two variants were however, found to be alike in their vegetative and floral morphology indicating themselves as minor infra-specific categories like chemotypes as reported earlier in species like *C. sulphuratum* (Baruah *et al.* 1999), *Litsea cubeba* (Baruah 2002, Nath *et al.* 1994, 1996) etc.

A key to the variants of *C. parthenoxylon* growing in Northeast India, formulated on the basis of their distinguishable characters is given below-

1a. Leaves and stem bark smelling like bottle brass leaf; areoles are comparatively smaller in size, average frequency of areole/mm^2 is 18.40, absolute areoles number in thousand is 54.60; major component in leaf and stem bark oils is 1, 8-cineole ....Variant I (1, 8-cineole chemotype).

1b. Leaves and stem bark smelling like linalool type-sweet basil; areoles are comparatively larger in size, average frequency of areole/mm^2 is 10.67, absolute areoles number in thousand is 39.94; major component in leaf and stem bark oils is linalool ....Variant II (Linalool chemotype).

Natural occurrence of 1, 8-cineole and linalool chemotypes of *C. parthenoxylon* are reported elsewhere (Zhu *et al.* 1994) other than India. The finding thus represents the first report on the occurrence of these two chemotypes of this species native to India (Baruah & Nath 2000).

The sources of essential oils rich in aroma components like 1, 8-cineole and linalool are of industrial importance. 1, 8-cineole is widely used in pharmaceutical preparations, to be applied internally and locally. Internally, it serves as a stimulating expectorant in cases of chronic bronchitis, while locally it is a mild anesthetic and antiseptic used in the treatment of inflammatory conditions. It is also used as room sprays, lotions and in all kinds of cosmetic preparations. On the other hand, linalool is a value added aroma chemical extensively used in perfume and flavor industries. It has marked effect on CNS including hypnotic, anticonvulsant and hypothermic activities.

As regards the aroma component like linalool in particular, the present production of this particular product from natural sources in India is insufficient even for internal consumption. Commercial cultivation of additional or alternative source of linalool in India is therefore needed to meet the industrial demand. Thus, Variant II of *C. parthenoxylon* leaves yielding 5% oil (FWB) with 86.90% linalool of the total oil may therefore be contemplated as potential source for the industry.

REFERENCES

Baruah A. 2000. *Cinnamomum* species associated with the livelihood of people in Northeast India: A systematic census with emphasis to ethnobotany. Ph. D. Thesis. Gauhati University. Guwahati. Assam.

Baruah A. 2002. *Litsea cubeba* Pers. A promising plant of industrial value. *Frag. Flav. Assoc. India J.* 4(3): 29-32.

Baruah A. and Nath S. C. 2000. Morphology and essential oils of two *Cinnamomum parthenoxylon* Meissn. variants growing in North-East India. *J. Med. Arom.* Pl. Sci. 22(1B): 370-376.

Baruah A., Nath S .C. and Boissya C .L. 1999. Taxonomic discrimination amongst certain chemotypes of *Cinnamomum sulphuratum* Nees. with emphasis to foliar micromorphology. *J. Swamy Botanical Club.* 16: 3-7.

Baruah A., Nath S. C. and Boissya C. L. 2000. Comparative foliar micromorphology of two chemotypes of *Cinnamomum parthenoxylon* Meissn. *J. Swamy Botanical Club.* 17(1&2): 25-28.

Hooker J. D. 1885. *Flora of British India.* Made and Printed by offset in the Great Britain by Willium Claws and Son's Ltd., London and Beccles. Vol 5.

Kanjilal U.N., Kanjilal P. C., De R. N., and Das A. 1940. *Flora of Assam.* Govt. of Assam Publication. Vol 4.

Kurz S. 1877. *Forest Flora of British Burma.* Published by order of the Govt. of India. Calcutta. Vol 2.

Nath S. C., Hzarika A. K. , Baruah A. and Sarma K. K. 1996. Essential oils of *Litsea cubeba* Pers – an additional chemotype of potential industrial value from North Eastern India. *J. Essential Oil Res.* 8(5): 575-576.

Nath S. C., Hzarika A. K., Singh R. S. and Ghosh A. C. 1994. *Litsea cubeba*: promising source of citral from Northeast India. *Indian Perfumer.* 38(1): 26-28.

Zhu L., Ding D., Lawrence B. M. 1994. The *Cinnamomum* species in China: Resources for the present and future. *Perfumer and Flavorist.* 19: 17-22.

18

Medicinal Plants: Potential Source in the Treatment of Diabetes

S. Majaw and D. Syiem

1. Introduction

Diabetes is defined as a state in which homeostasis of carbohydrate and lipid metabolism is improperly regulated by insulin. This results primarily in elevated fasting and postprandial blood glucose levels. If this imbalanced homeostasis continues for a protracted period of time, it leads to hyperglycemia that in due course turns into a syndrome called diabetes mellitus which is broadly classified into two types: Type 1 diabetes mellitus also called insulin-dependent diabetes mellitus (IDDM) and Type 2 or the non-insulin dependent diabetes mellitus (NIDDM). It is estimated that incidence of Type 1 diabetes will be about 40% higher in the year 2010 than in 1997 and yet there is no identified agent substantially capable of preventing this type of disease (Schatz *et al.* 2000; Atkinson & Eisenbarth, 2001). Type 2 or NIDDM is far more common and results from a combination of defects in insulin secretion and action. Type 2 accounts for 90 to 95% of all diabetic patients. Treatment of Type 2 diabetes is complicated by several factors inherent to the disease process, typically, insulin resistance, hyperinsulinemia, impaired insulin secretion, and reduced insulin-mediated glucose uptake and utilization (De Fronzo *et al.* 1997; Polonsky, 1996; Groop, 1989).

Long-standing diabetes mellitus is associated with an increased prevalence of microvascular and macrovascular complications. With the rising prevalence of diabetes, the number suffering from the vascular complications of diabetes will also increase. As the disease progresses, tissue or vascular damage develops leading to severe diabetic complications such as diabetic cataract (Jansirani & Anathanaryanan, 2004), retinopathy

(Klein, 2004), neuropathy (Vinik *et al.*, 2004), nephropathy (DeFronzo, 2004), cardiovascular complications (Young & Chyun, 2004) and ulceration (Coleman, 2004). Thus, diabetes covers a wide range of heterogeneous pathophysiological conditions. Pathways and mechanisms implicated with diabetes complications include: Polyol pathway, formation of Advanced Glycation End products (AGEs), oxidative / carbonyl stress, activation of protein kinase C (PKC), increased flux through the hexosamine biosynthesis pathway *(HBP)* (Brownlee, 2001).

Aldose reductase (EC 1.1.1.21) is the first enzyme in the polyol pathway. It is a cytosolic, monomeric oxidoreductase that catalyses the NADPH-dependent reduction of a wide variety of carbonyl compounds, including glucose. Aldose reductase has a low affinity (high K_m) for glucose and at normal glucose concentrations that is found in non-diabetics, metabolism of glucose by this pathway is a very small percentage of total glucose use. But in a hyperglycaemic environment, increased intracellular glucose results in its increased enzymatic conversion to the polyalcohol sorbitol with concomitant decrease in NADPH. In the polyol pathway, sorbitol is further oxidized to fructose by the enzyme sorbitol dehydrogenase (SDH) (EC 1.1.1.14), with NAD^+ reduced to NADH. Flux through this pathway during hyperglycaemia varies from 33% of total glucose in the rabbit lens to 11% in human erythrocytes. Thus, the contribution of this pathway to diabetic complications may be very much species specific and tissue dependent (Brownlee, 2001).

A number of mechanisms have been proposed to explain the potential detrimental effects of hyperglycaemia-induced that increase in polyol pathway flux. These include sorbitol-induced osmotic stress, decreased (Na^+ + K^+) ATPase activity, an increase in cytosolic NADH/NAD^+ and a decrease in cytosolic NADPH. Sorbitol does not diffuse easily across cell membranes, its accumulation results in osmotic damage to microvascular cells (Brownlee, 2001).

Although several drugs targeted for carbohydrate hydrolyzing enzymes (pseudosaccharides), release of insulin from pancreatic â-cells (sulphonyl urea), glucose utilization (biguanides), insulin sensitizers, PPARá agonists (glitazones) are in clinical practice, the growing diabetes market observes a number of changes. The glitazones are meant to target the problem of insulin resistance and enhance insulin action at the cellular level; however, some of these drugs (troglitazone) are linked to liver toxicity, including a number of deaths from hepatic failure (Krische, 2000; Gale, 2001; Stern, 1992) and raising the symptoms and risk factors of heart disease leading to heart failure (Gale, 2001). Therefore, the long term effect and risk of these drugs are not yet clear. On the other hand, traditional medicinal plants with various properties have been used since ancient times by physicians and laymen to treat a great variety of human diseases including diabetes, coronary heart disease and cancer (Middleton *et al.* 2000; Havsteen 1984). Many plants have been reported for the anti-hyperglycemic activity (Marles & Farnsworth, 1995). The beneficial multiple activities like manipulating carbohydrate metabolism by various mechanisms, preventing and restoring integrity and function of b-cells, insulin-releasing activity, improving glucose uptake and utilization and the antioxidant properties present in medicinal plants offer exciting opportunity to develop them into novel therapeutics. The multiple activities of plant-based medicinal preparations meant for diabetes offer enormous scope for combating the threat of the diabetic epidemic. It is important therefore, that herbal medicines and preparations, with its holistic therapeutic approach should be given serious attention.

Earlier it was reported that roots of *Potentilla fulgens* L. (Rosaceae), *Osbeckia chinensis* L. (Melastomaceae), are common plants of the Northeastern Region of India and used by local tribes, exhibit hypoglycemic and anti-hyperglycemic properties (Syiem *et al.* 2002; Syiem & Khup, 2006). This paper present an initial finding on our study to see the effects of *P. fulgens*, *O. chinensis*, *Clerodendron celebrookianum* and *Areca catechu* on the sorbitol dehydrogenase involve in the polyol pathway. *C. celebrookianum* Walp. (Verbenaceae) leaves are used as a decoction for hypertension (Lalramnghinglova, 2001) and seeds of *A. catechu* L. (Palmaceae) have been studied for its antioxidant property (Wetwitayaklung *et al.*, 2006).

2. Materials and Methods

Plant Material

In the present study, the SDH inhibitory activity of 4 different plant extracts were studied. The plants were acquired from different locations of Meghalaya. Methanolic extracts were prepared according to the method of Harborne (1998).

Test animals

Healthy, adult Swiss albino mice of approximately the same age group and weighing 20–30 g were used for the study. Mice were housed in a room kept under controlled conditions with temperature maintained at 22°C on a 12-h light/dark cycle and were fed with balanced mice feed obtained from Amrut Laboratory (Pune, India).

Bioassay

The bioassay procedure with some modifications was followed according to the method of Haraguchi *et al.* (1997).

3. Results

Table 1 and Table 2, showed the percentage inhibition values of sorbitol dehydrogenase by 4 extracts in liver and kidney of mice respectively. The inhibitory activity of the extracts was calculated using the following formula,

$$\% \text{ SDH} = \frac{\Delta \text{Abs. (negative control)} - \Delta \text{Abs. (Extract)} \times 100}{\Delta \text{Abs. (negative Control)}}$$

Where, control (liver/kidney) and diabetic (liver/ kidney) without extract was taken as the negative control.

Table 1

Effect of extract on Sorbitol Dehydrogenase activity of mice liver. Values are means ± SD. Sample size is 6 mice

Sl. No.	Plants	SDH % inhibition Control Mice	SDH % inhibition Diabetic Mice
1	*Potentilla fulgens*	72.51% ± 0.58	88.45% ± 0.42
2	*Areca catechu*	43.29% ± 0.74	75.55% ± 0.69
3	*Osbeckia chinensis*	6.31% ± 0.16	16.88% ± 0.88
4	*Clerodendron celebrookianum*	0% ± 0.00	3.60% ± 0.33

Table 2

Effect of extract on Sorbitol Dehydrogenase activity of mice kidney. Values are means ± SD. Sample size is 6 mice

Sl. No.	Plants	SDH % inhibition Control Mice	SDH % inhibition Diabetic Mice
1	*Potentilla fulgens*	76.56% ± 0.26	83.01% ± 0.66
2	*Areca catechu*	62.00% ± 0.72	68.64% ± 0.52
3	*Osbeckia chinensis*	7.72% ± 0.34	17.83 ± 0.18
4	*Clerodendron celebrookianum*	0% ± 0.00	6.69% ± 0.34

4. Discussion

Among the four evaluated plants, *P. fulgens* showed maximum SDH% inhibition in diabetic liver (88.45%) when compared to *A. catechu* (75.55%), *O. chinensis* (16.88%) and *C. celebrookianum* (3.60%). Similarly, in diabetic kidney *P. fulgens* showed maximum inhibition (83.01%) in comparison to *A. catechu* (68.64%), *O. chinensis* (17.83%) and *C.*

celebrookianum (6.69%). The magnitude of inhibition of extract on the enzyme was found to be higher in diabetic mice compared to that of the control. Similar results have been reported with *Scoparia dulcis* L. (Latha & Pari, 2004), *Cucurma longa* L. (Arun & Nalini 2002) and *Trigonella foenum graecum* (Preet *et al.* 2005). In addition, many plants known to inhibit aldose reductase, the other enzyme of the polyol pathway, have also been reported and reviewed which include *Cinnamomum cassia* (Lee, 2002), *Myrciaria dubia, Monochasma savatierii* and *Perilla frutescens* (Guzmán & Guerrero, 2005).

5. Conclusion

Earlier it was reported that roots of *Potentilla fulgens* L. (Rosaceae) possess hypoglycemic and anti-hyperglycemic properties (Syiem *et al.* 2002). The goal of this study was to evaluate the effect of *P. fulgens* methanolic extract on the polyol pathway enzymes. Our data show that, the extract could effectively inhibit the sorbitol dehydrogenase enzyme in diabetic mice. Therefore, *P. fulgens* presents a potential plant for the treatment of diabetes as well as its associated complications.

REFERENCES

Arun N. and Nalini N. 2002. Efficacy of turmeric on blood sugar and polyol pathway in diabetic albino rats. *Plant Foods Hum Nutr* 57: 41-52.

Atkinson M.A. and Eisenbarth G.S. 2001. Type 1 diabetes: new perspectives on disease pathogenesis and treatment. *Lancet* 358: 221–229.

Brownlee M. 2001. Review article: Biochemistry and molecular cell biology of diabetic complications. *Nature* 414: 813-820.

Coleman W.C. 2004. The diabetic foot. In Inzucchi SE, eds., *The Diabetes Mellitus Manual*, McGraw Hill, USA, pp. 429-438.

DeFronzo R.A. 1997. Pathogenesis of type 2 diabetes: metabolic and molecular implications for identifying diabetes genes. *Diabetes Rev* 5: 177–269.

DeFronzo R.A. 2004. Diabetic Nephropathy. In. Inzucchi SE, eds., *The Diabetes Mellitus Manual*, Mc Graw Hill, USA, pp. 325-346.

Gale E.A.M. 2001. Lessons from the glitazones: a story of drug development. *Lancet* 357: 1870–1875.

Groop L.C., Bonadonna R.C., Del Prato S., Ratheiser K., Zick K., Ferranini E. and De Fronzo R.A. 1989. Glucose and free-fatty acid metabolism in non-insulin-dependent diabetes: evidence for multiple insulin resistance. *J Clin Invest* 84: 205-213.

Guzmán Á. and Guerrero R.O. 2005. Inhibition of aldose reductase by herbs extracts and natural substances and their role in prevention of cataracts. *Rev Cubana Plant Med* 10: 3-4

Haraguchi Ohmi I., Fukuda A., Tamura Y., Mizutani K. and Tanaka O. 1997. Inhibition of aldose reductase and sorbitol accumulation by astibin and taxifolin dihydroflavonols in *Engelhardtia chrysolepsis*. *Biosci Biotechnol Biochem* 61:651-654.

Harborne J.B. 1998. *Phytochemical Methods*. 3rd ed. London, Chapman & Hall, pp. 4–7.

Havsteen B. 1984. Flavonoids, a class of natural products of high pharmacological potency. *Biochem Pharmacol* 33:1367–1369.

Jansirani P. and Anathanarayanan H. 2004. A comparative study of lens protein glycation in various forms of cataract. *Indian J Clin Biochem* 19: 110-112.

Klein R. 2004. Retinopathy and other Ocular complications in Diabetes. In. Inzucchi SE, eds., *The Diabetes Mellitus Manual*, Mc Graw Hill, USA, pp. 303-324.

Krische D. 2000. The glitazones: proceed with caution. *West J Med* 173:54-57.

Lalramnghinglova H. 2001. In: *Ethnomedicine of North East India*. Proceedings of National Seminar on Traditional knowledge Base of Herbal Medicines and Plant Resources of North East India, Protection Utilization and Conservation: CSIR, New Delhi.

Latha M. and Pari L. 2004. Effect of an aqueous extract of *Scoparia dulcis* on blood glucose, plasma insulin and some polyol pathway enzymes in experimental rat diabetes. *Braz J Med Biol Res.* 37: 577-586.

Lee H.S. 2002. Inhibitory activity of Cinnamomum cassia bark-derived component against rat lens aldose reductase. *J Pharm Pharmaceut Sci* 5:226-230.

Marles R.J. Farnsworth N.R. 1995. Anti-diabetic plants and their active constituents. *Phytomedicine* 2: 137–189.

Middleton E.Jr., Kandaswami C. and Theoharides T.C. 2000. The effect of plant Flavonoids on mammalian cells. *Pharmacol Rev* 52: 673–751.

Polonsky K.S., Sturis J. and Bell G.I. 1996. Seminars in Medicine of the Beth Israel. *N Engl J Med* 334: 777–783.

Preet A., Gupta B.L., Siddiqui M.R., Yadava P.K. and Baquer N.Z. 2005. Restoration of ultrastructural and biochemical changes in alloxan-induced diabetic rat sciatic nerve on treatment with Na_3VO_4 and *Trigonella*—a promising antidiabetic agent. *Mol Cel Biochem* 278: 21-31.

Schatz D., Krischer J. and Skyler J. 2000. Now is the time to prevent type 1 diabetes. *J Clin Endocrinol Metab* 85: 494–522.

Stern M.P. 1999. On the need for outcome trials in preventive pharmacology: lessons from the recent. *Diabetes Care* 22: 844–845.

Syiem D., Gareth S., Khup P.Z., Khongwir B.S., Kharbuli B. and Kayang H. 2002. Hypoglycemic effect of *Potentilla fulgens* L. in normal and alloxan-induced diabetic mice. *J Ethnopharmacol* 83: 55–56.

Syiem D. and Khup P.Z. 2006. Study of the Traditionally Used Medicinal Plant *Osbeckia chinensis* for Hypoglycemic and Anti-hyperglycemic Effects in Mice. *Pharmacol Biol* 44:613-618,

Vinik A.I., Erbas T., Pfeifer M.A., Feldman E.L., Stevens M.J. and Russell J.W. 2004. Diabetic Automic Nephropathy. In. Inzucchi SE, eds., *The Diabetes Mellitus Manual*, McGraw Hill,USA, pp. 347-365.

Young L.H. and Chyun D.A. 2004. Heart Disease in patients with Diabetes In. Inzucchi SE, eds., *The Diabetes Mellitus Manual*, McGraw Hill, USA, pp. 408-420.

Wetwitayaklung P., Phaechamud T., Limmatvapirar C. and Keokitichai S. 2006. The study of antioxidant capacity in various parts of Areca catechu L. *Naresuan Univ J* 14: 1-14.

19

Probiotics: Indigenous Fermented Foods as a Source of Potential Medicinal Microbes

Rakshak Kumar and S. R. Joshi

1. Introduction

Almost 100 years ago, the Russian born bacteriologist and 1908 Nobel Laureate, Ilya Mechnikov stated, "Death begins in the colon". In his ground-breaking book *Prolongation of Life: Optimistic Studies*, Mechnikov documented in detail the unusually lengthy life spans of several societies that practiced eating fermented foods and special bacterial cultures called kefirs. He made consistent observations of 100+ years old men and women that were still leading highly active and healthy lives. His research made a direct link between human longevity and the necessity of maintaining a healthy balance of probiotics in the body. Hence, the concept of probiotics evolved since then (Metchnikoff E. 1907). However, the term probiotics is taken from Greek and literally means 'for life' and was only coined in 1965 by Lilley and Stillwell to refer to the stimulation of the growth of one microbe by another (Lilley D. M. & Stillwell R. H., 1965), (the opposite of antibiotics). Later a British scientist, Roy Fuller popularized its definition in 1989, defining it as "a live microbial feed supplement, which beneficially affects the host animal by improving its intestinal balance" (Fuller R. 1989). Hence, probiotics are nothing but viable nonpathogenic microbes that when ingested exert a positive influence on the health or physiology of the consumer and probiotic food are food containing these live microbes believe to actively enhance the health of the host by improving the balance of micro-flora in the gut.

The dietary use of live microorganisms has a long history. Mention of cultured dairy products is found in the Bible and the sacred books of Hinduism. Soured milks and cultured dairy products, such as kefir, koumiss, leben and dahi were often used therapeutically before the existence of microorganisms was recognized. The use of microorganisms in food fermentation is one of the oldest methods for producing and preserving food. Much of the world depends upon various fermented foods that are staples in the diet (Joshi S. R. 2007). These fermented foods are natural habitat of the probiotic microorganisms. In other way, those microbes, which help in the fermentation of these traditional foods, can have the potential probiotic effect. Shahani & Ayebo (1980) reported that potential health benefits might result from growth and action of the bacteria during the manufacture of cultured foods. The microflora of fermented food are capable of providing numerous health benefits beyond basic nutritional value (Alm L. 1982, Hosono A. 1986, Benchimol E.I. & Mack D.R. 2004).

Probiotic microorganisms that can have positive health effects when ingested are increasingly studied and used in humans. They have shown particular promise in the treatment of acute infectious diarrhea and the prevention of antibiotic-associated diarrhea, associations that have been investigated widely and have been subject to systematic review (Benchimol E.I. & Mack D.R. 2004, Van Niel C.W., Feudtner C., Garrison M.M., Christakis D.A. 2002, Cremonini F., Di Caro S., Nista E.C., *et al.* 2002, D'Souza A.L., Rajkumar C., Cooke J., Bulpitt C.J. 2002). Other conditions potentially treatable by probiotics include chronic diarrhea (Xiao S.D., Zhang de Z., Lu H. 2003), inflammatory bowel disease (Schultz M., Timmer A., Herfarth H.H., Sartor R.B., Vanderhoof J.A., Rath H.C. 2004), irritable bowel syndrome (Saggioro A. 2004) and food allergy (Majamaa H., Isolauri E. 1997), potentially preventable conditions range from traveler's diarrhea (Hilton E., Kolakowski P., Singer C., Smith M. 1997) and necrotizing enterocolitis (Dani C., Biadaioli R., Bertini G., Martelli E., Rubaltelli F.F. 2002) to urogenital infections (Cadieux P., Burton J., Gardiner G. 2002), atopic diseases (Kalliomaki M., Salminen S., Arvilommi H., Kero P., Koskinen P., Isolauri E. 2001), control of serum cholesterol levels (Lin S.Y., Ayres J.W., Winkler W. and Sandine W.E. 1989), Wan-Kyu Lee, Hyeong-Jun Lim, So-Young Kim, Hiromi Kimoto, Sadahiro Ohmomo, Yashuhito Tashiro & Hideki Takebe (2004), cancer (Gilliland S.E. 1990, O'Sullivan M.G., Thornton G., O'Sullivan G.C. & Collins J.K. 1992) and dental caries (Nase L., Hatakka K., Savilahti E. 2001). Vanderhoof (Vanderhoof J.A. 2001) and other leaders in the field propose future applications in cystic fibrosis, rheumatoid arthritis, and cancers. Figure 1 indicates the summary of health benefits from probiotics (S. Parvez, K.A. Malik, S. Ah Kang & H -Y. Kim. 2006).

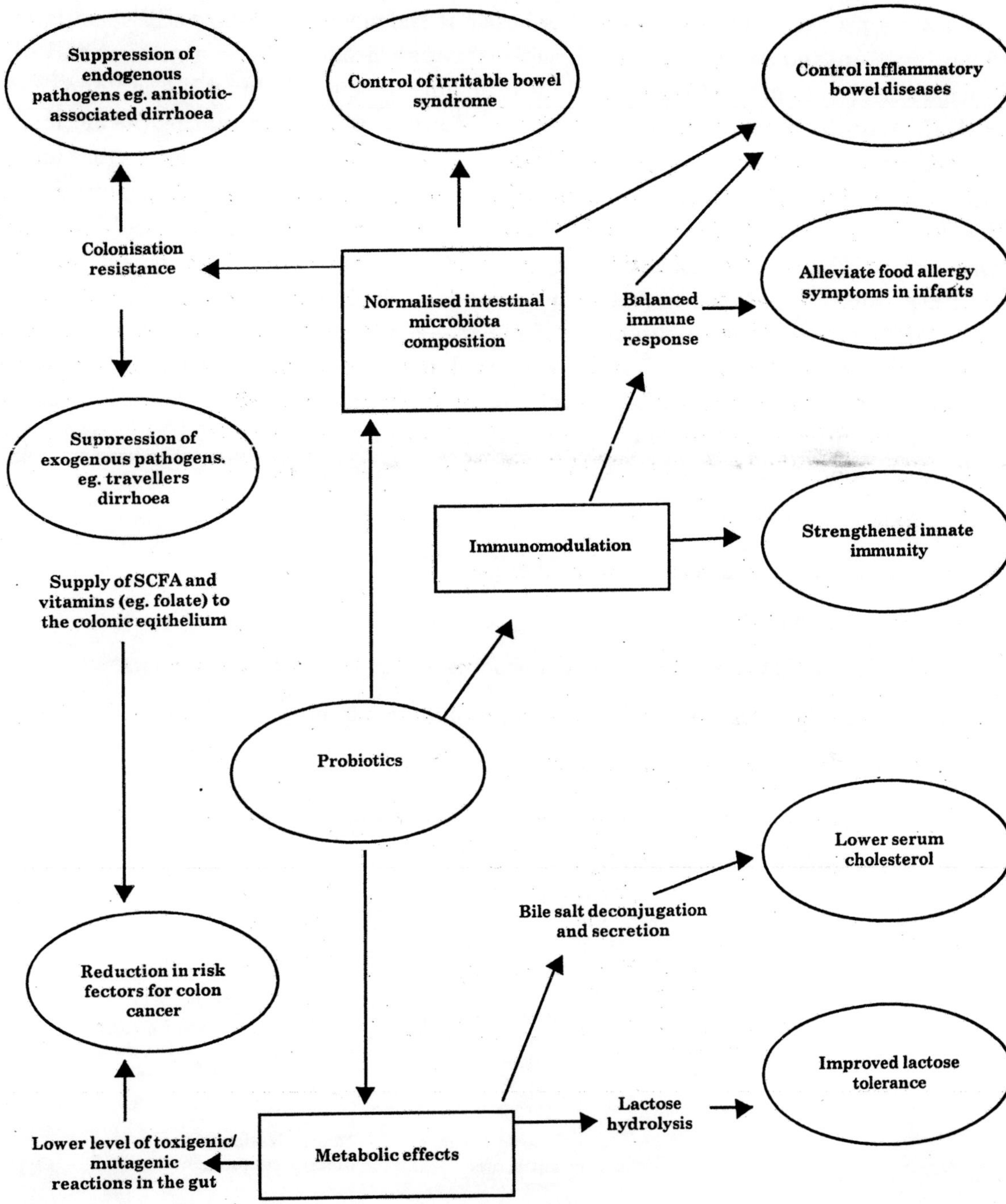

Figure 1. Various health benefits from probiotics consumption

Probiotics exert their effects on the host but the mechanisms are still speculative. They may antagonize pathogens directly through production of antimicrobial and antibacterial compounds such as cytokines and butyric acid (De Vuyst L. and Vandamme E.J. 1994, Kailasapathy K. & Chin J. 2000), reduce gut pH by stimulating the lactic acid to produce microflora (Langhendries J.P., Detry J., Van Hees J. 1995), compete for binding and receptor sites that pathogens occupy (Kailasapathy K. & Chin J. 2000, Fujiwara S., Hashiba H., Hirota T., Forstner J.F. 1997), improve immune function and stimulate immunomodulatory cells (Isolauri E., Juntunen M., Rautanen T., Sillanaukee P., Koivula T. 1991, Isolauri E., Joensuu J., Suomalainen H., Luomala M., Vesikari T. 1995, Rolfe R.D. 2000), produce lactase which aids in lactose digestion, compete the nutrients and adhere the sites on the gut wall and regulate colonocyte gene expression (e.g. expression of mucin genes) (Fooks L.J. & Gibson G.R. 2002, Mack D.R. 1999, Steer T. 2000). A group of requirements have been identified for a microorganism to be defined as an effective probiotic (Salminen S., Isolauri E., Salminen E. 1996). They are-

1. Adherence to cells.
2. Exclude or reduce pathogenic adherence.
3. Persist and multiply.
4. Produce acids, peroxide, and bacteriocins antagonistic to pathogen growth.
5. They are safe, non-invasive, non-carcinogenic and nonpathogenic.
6. Co-aggregate to form a normal balanced flora.

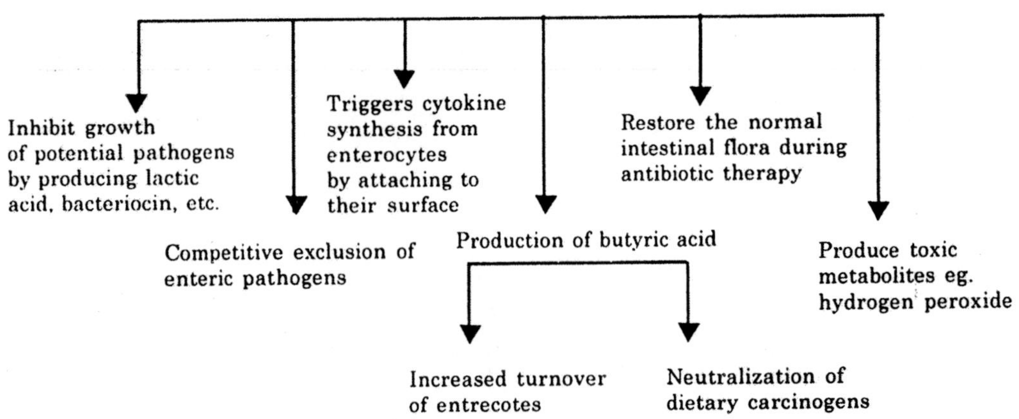

Figure 2. The Suggested mechanisms underlying probiotic activity (Kaur N. & Gupta A.K. 2002)

Traditional fermented foods and probiotics

With regards to the traditional fermented foods and their relevance to probiotic microorganisms, mention can be made of *tungrymbai* and *tungtap*, which are the traditional fermented food products prepared from soybean and fish respectively in Meghalaya by the indigenous Khasi tribes. Fermented foods are typically of the region and exhibit unique flavours and textures that may not be palatable to everyone. *Tungrymbai* is a popular fermented soybean based sticky food and *tungtap* is a popular fermented dry fish product that serves as a cheap source of high protein food in local diet. Preparation and consumption of these foods reflect deep-rooted food culture of the ethnic tribes. People using indigenous knowledge exclusively practice preparation of *tungrymbai* and *tungtap*. *Tungrymbai* is prepared using soaked soybeans, which are cooked until they can be pressed easily. Excess water is drained off and placed in basket lined with locally grown fresh leaves of *Pyrnium pubinerve* Bl. (Fam: Marantaceae; locally called *slamet*) and jute bags and left to ferment naturally in ambient temperature preferably near/over the fire place (25-40°C) for 3-4 days. It is similar to *kinema* of Bhutan, Nepal, Darjeeling and Sikkim in India. It is reported to be a good source of protein and other nutrients (Agrahar-Murugkar D. & Subhulakshmi G. 2006). *Tungtap* is prepared using fish (*Danio* spp) which is sun dried, mixed with salt, kept in an earthen pot and fermented for 4-7 days.

2. Material and Methods

The present study assesses the microbiota associated with natural fermentation of *tungrymbai* and *tungtap*; and then relates the microbes obtained from these indigenous fermented foods as the potential probiotic microbes by comparing it with microbes which already has been studied extensively and for which evidences are available, wherein clinical trials have been performed to prove their efficacy against various human or animal ailments.

Collection of samples: *Tungrymbai* and *tungtap* samples were purchased from local (Iewduh or Bara Bazar) market, aseptically kept in icebox and transported to laboratory for analysis.

Microbial analysis: 10 g of well mixed samples were blended in 90 ml of 0.85% (w/v) sterile physiological saline in a lab-blender for 10 min. Decimal dilution series was prepared in sterile diluent and 1 ml of appropriate diluted suspension was mixed with molten media and poured into plates. Lactic acid bacteria were enumerated on de Man, Rogosa and Sharpe (MRS) agar (HiMedia M641) supplemented with 1% calcium carbonate and incubated anaerobically in an Anaerobic Gas-Pack System (HiMedia LE002, India) at 30°C for 72 h. Total viable counts were determined using plate count agar (HiMedia M091A) after incubating at 30° C for 48 h. Presence of Enterobacteriaceae was tested by using selective violet red bile glucose agar (HiMedia M581) and incubating at 30° C for 48 h.

Samples were examined for the presence of yeasts and moulds, using malt extract agar (HiMedia M924) supplemented with 100mg L^{-1} chloramphenicol and incubating anaerobically at 28°C for 72 h.

Characterization and Identification

Colony, cell morphology and Gram staining comprise initial characterization of bacterial isolates. All other phenotypic characterizations of bacterial isolates were carried as per standard protocols (Gordon R.E., Haynes W.C. & Pang C.H-N. 1973, Schillinger U. & Lucke F.K. 1987). Endospore forming bacteria were identified according to the keys based on Claus D. & Berkeley R.C.W. (1986). Lactic acid bacteria were identified following the taxonomic keys laid down in Bergey's Manual of Systematic Bacteriology (Sneath P.H.A., Mair N.S., Sharpe M.E. & Holt J.G. 1986). Yeasts isolated were identified by the standard morphological and biochemical tests (Kurtzman C.P. & Fell J.W. 1998).

Comparison of the microbes found in tungrymbai and tungtap with the studied potential probiotics

The microbial population observed and identified from *tungrymbai* and *tungtap* were used as the keywords to search for their potential probiotic effect using the Internet based search engines like http://www.google.co.in/ and http://www.ncbi.nlm.nih.gov/.

3. Results

The microbial population in the fermentation of *tungrymbai* revealed the presence of strains of rod-shaped, gram positive, endospore forming bacteria identified as *Bacillus subtilis* according to the criteria laid down by Claus and Berkeley (Claus D. & Berkeley R.C.W. 1986). Strains of cocci-shaped, gram-positive, non-spore forming lactic acid bacteria were identified as *Lactococcus lactis*, *Lactococcus plantarum* and *Leuconostoc mesenteroides* following taxonomical keys laid down in Bergey's Manual of Systematic Bacteriology (Sneath P.H.A., Mair N.S., Sharpe M.E. & Holt J.G. 1986). Following the taxonomical keys (Kurtzman C.P. & Fell J.W. 1998), yeasts recovered from various sources were identified as *Candida parapsilosis, Saccharomyces bayanus, Saccharomycopsis fibuligera* and *Geotrichum candidum.*

The microbial population in the fermentation of *tungtap* revealed the presence of strains of cocci-shaped, gram-positive, non-spore forming lactic acid bacteria identified as *Lactococcus lactis, Lactococcus plantarum, Lactobacillus plantarum, Lactobacillus fructosum* and *Leuconostoc mesenteroides* following taxonomical keys laid down in Bergey's Manual of Systematic Bacteriology (Sneath P.H.A., Mair N.S., Sharpe M.E. & Holt J.G. 1986). Following the taxonomical keys (Kurtzman C.P. & Fell J.W. 1998), yeasts recovered from various sources were identified as *Saccharomyces cerevisiae, Geotrichum candidum, Aspergillus glaucus, Pediococcus cerevisiae, Candida* sp. and *Penicillium* sp.

Search result fetched a number of research papers wherein clinical trials have been performed using animal models to check the efficacy of these microbes against various human or animal ailments. Search results, comprising the name of microbes, the source from which we obtained them, their associated probiotic effect and the referred papers are shown in Table 1.

Table 1

Microbial population of *tungrymbai* and *tungtap*, and their comparison with the type microorganism which has potential probiotic effects, supported with the studied research papers

Microflora	Sources	Associated Probiotic effects	Reference paper
Bacillus subtilis	*Tungrymbai*	Production of bacteriocins against Non-Lactic Microorganisms.	Naidu A.S. *et al.* (1999)
		Colonization and antimicrobial activity	Pinchuk I.V. *et al.* (2001)
Lactococcus lactis	*Tungrymbai, Tungtap*	Decreases total serum cholesterol and triglycerides, hence can be effective in prevention of hypercholesterolemia and fatty liver.	Lee W.K *et al.* (2004)
		Potential as antigen presenting vehicles suitable for mucosal administration	Mercenier, (1999); Wells *et al.*, (1995),(1996)
Lactococcus plantarum	*Tungrymbai, Tungtap*	No reports	
Leuconostoc mesenteroides	*Tungrymbai Tungtap*	No reports	

(Contd...)

Microflora	Sources	Associated Probiotic effects	Reference paper
Lactobacillus plantarum	*Tungtap*	Reduced incidence of diarrhoea in daycare centres when administered to only half of the children	Vanderhoof (2000)
		Especially effective in reducing inflammation in inflammatory bowel;	Schultz and Sartor (2000); Vanderhoof (2000)
		e.g., enterocolitis in rats, small bowel bacterial overgrowth in children, pouchitis	Vanderhoof (2000)
		Reduced pain and constipation of irritable bowel syndrome	Nobaek *et al*. (2000)
		Reduced bloating, flatulence, and pain in irritable bowel syndrome in controlled trial.	Walker (2000)
		Positive effect on immunity in HIV+ children	
Lactobacillus fructosum	*Tungtap*	No reports	
Saccharomyces cerevisiae	*Tungtap*	Fewer relapses of inflammatory bowel disease	Guandalini *et al*. 2000

REFERENCES

Agrahar-Murugkar D and Subhulakshmi G. 2006. Preparation techniques and nutritive value of fermented foods from the Khasi tribes of Meghalaya. *Ecology of Food and Nutrition*, 45 (1): 27-38

Alm L. 1982. Effect of fermentation on lactose, glucose and galactose content in milk and suitability of fermented milk products for lactose-deficient individuals. *Journal of Dairy Science*, 65: 346–352.

Benchimol E.I. and Mack D.R. 2004. Probiotics in relapsing and chronic diarrhea. Journal of Pediatric Hematology Oncology, 26: 515– 517.

Cadieux P., Burton J., Gardiner G. 2002. Lactobacillus strains and vaginal ecology. *Journal of the American Medical Association*, 287: 1940–1941.

Claus D. and Berkeley R.C.W. 1986. Genus Bacillus Cohn 1872,174. In: *Bergey's Manual of Systematic Bacteriology.* Vol. 2 (Sneath P.H.A., Mair N.S., Sharpe M.E. and Holt J.G. eds.), Williams and Wilkins, Baltimore, pp 1105-1139.

Cremonini F., Di Caro S., Nista E.C., *et al.* 2002. Meta-analysis: the effect of probiotic administration on antibiotic-associated diarrhoea. *Aliment Pharmacology Therapeutic,* 16:1461–1467

Dani C., Biadaioli R., Bertini G., Martelli E., Rubaltelli F.F. 2002. Probiotics feeding in prevention of urinary tract infection, bacterial sepsis and necrotizing enterocolitis in preterm infants. A prospective double-blind study. *Biology of the* Neonate, 82: 103– 108.

De Vuyst L. and Vandamme E.J. 1994. Antimicrobial potential of lactic acid bacteria. In: De Vuyst L., Vandamme E.L., (Eds). *Bacteriocins of Lactic acid Bacteria.* Glasgow, United Kingdom: Blackie Academic and Professional, p: 91-142.

D'Souza A.L, Rajkumar C., Cooke J., Bulpitt C.J. 2002. Probiotics in prevention of antibiotic associated diarrhoea: meta-analysis. *British Medical Journal,* 324: 1361.

Fooks L.J. and Gibson G.R. 2002. Probiotics as modulators of the gut flora. *British Journal of Nutrition,* 88: 39-49.

Fujiwara S., Hashiba H., Hirota T., Forstner J.F. 1997. Proteinaceous factor(s) in culture supernatant fluids of bifidobacteria which prevents the binding of enterotoxigenic *Escherichia coli* to gangliotetraosylceramide. *Applied and Environmental Microbiology,* 63: 506- 512.

Fuller R. 1989. Probiotics in man and animals. *Journal of Applied Bacteriology,* 66: 365– 378.

Gilliland S.E. 1990. Health and Nutritional benefits from lactic acid bacteria. *FEMS, Microbiological Reviews,* 87: 175–188.

Gordon R.E., Haynes W.C. and Pang CH-N. 1973. *The Genus Bacillus,* Handbook No. 427. United States Department of Agriculture, Washington D.C., p.97.

Guandalini S., Pensabene L., Zikri M.A., Dias J.A., Casali L.G., Hoekstra H., Kolacek S., Massar K., Micetic-Turk D., Papadopoulou A., de Sousa J.S., Sandhu B., Szajewska H. and Weizman Z. 2000. *Lactobacillus* GG administered in oral rehydration solution to children with acute diarrhea: a multicenter European trial. Journal of Pediatric Gastroenterology Nutrition, 30: 54–60.

Hilton E., Kolakowski P., Singer C., Smith M. 1997. Efficacy of *Lactobacillus* G.G. as a diarrheal preventive in travelers. *Journal of Travel Medicine,* 4: 41–43.

Hosono A. 1986. Anti-mutagenic properties of lactic-acid cultured milk on chemical and fecal mutagens. *Journal of Dairy Science,* 69: 2237–2242.

Isolauri E., Juntunen M., Rautanen T., Sillanaukee P., Koivula T. 1991. A human Lactobacillus strain (Lactobacillus casei sp. Strain GG) promotes recovery from acute diarrhea in children. *Pediatrics,* 88: 90-7.

Isolauri E., Joensuu J., Suomalainen H., Luomala M., Vesikari T. 1995. Improved immunogenicity of oral D 3 RRV reassortant rotavirus vaccine by Lactobacillus casei GG. *Vaccine,* 13: 310-2.

Joshi S.R. 2007. *Microbes: Redefined Personality.* A P H Publishing Corporation, New Delhi. p.97-112.

Kailasapathy K. and Chin J. 2000. Survival and therapeutic potential of probiotic organisms with reference to Lactobacillus acidophilus and *Bifidobacterium* spp. *Immunology and Cell Biology,* 78: 80-88.

Kalliomaki M., Salminen S., Arvilommi H., Kero P., Koskinen P., Isolauri E. 2001. Probiotics in primary prevention of atopic disease: a randomized placebo-controlled trial. *Lancet.* 357: 1076-1079.

Kaur N. and Gupta A.K. 2002. Applications of inulin and oligofructose in health and nutrition. *Journal of Bioscience.* 27: 703–714.

Kurtzman C.P. and Fell J.W. 1998. *The Yeasts, a Taxonomic Study.* Elsevier Science Publishers, Amsterdam, 4th Ed, p 1074.

Langhendries J.P., Detry J., Van Hees J. 1995. Effect of a fermented infant formula containing viable bifidobacteria on the fecal flora composition and pH of healthy full-term infants. *Journal of Pediatric and Gastroenterology Nutrition,* 21: 177-81.

Lilley D. M., and Stillwell R. H. 1965. Probiotics: growth promoting factors produced by microorganisms. *Science,* 147: 747–748.

Lin S.Y., Ayres J.W., Winkler, W. and Sandine W.E. 1989. Lactobacillus effects on cholesterol: in vitro and in vivo results. *Journal of Dairy Research,* 72: 2885–2889.

Mack D.R. 1999. Probiotics inhibit enteropathogenic *E. coli* adherence *in vitro* by inducing intestinal mucin gene expression. *American Journal of Physiology,* 276: 941-950.

Majamaa H., Isolauri E. 1997. Probiotics: a novel approach in the management of food allergy. *Journal of Allergy & Clinical Immunology,* 99: 179–185.

Mercenier A. 1999. Lactic acid bacteria as live vaccines. In: *Probiotics: A Critical Review.* G.W.Tannock, ed. Horizon Scientific Press, Wymondham, UK. p. 113-127.

Metchnikoff, E. 1907. *The Prolongation of Life. Optimistic Studies.* William Heinemann, London, United Kingdom.

Naidu A. S., Bidlack W. R. and Clemens R. A. 1999. Probiotic Spectra of Lactic Acid Bacteria (LAB) *Critical Reviews in Food Science and Nutrition,* 38(1): 13–126

Nase L., Hatakka K., Savilahti E. 2001. Effect of long-term consumption of a probiotic bacterium, *Lactobacillus rhamnosus* GG, in milk on dental caries and caries risk in children. *Caries Research.* 35: 412–420.

Nobaek S., Johansson, M.L. and Molin, G. 2000 Alteration of intestinal microflora is associated with reduction in abdominal bloating and pain in patients with irritable bowel syndrome. *American Journal of Gastroenterology,* 95: 1231–1238.

O'Sullivan M.G., Thornton G., O'Sullivan G.C. and Collins, J.K. 1992. Probiotic bacteria: myth or reality. *Trends in Food and Science Technology,* 3: 309–314.

Pinchuk I. V., Bressollier P., Verneuil B., Fenet B., Sorokulova I. B., Megraud F. and Urdaci M. C. 2001. *In vitro* anti-*Helicobacter pylori* activity of the probiotic strain *Bacillus subtilis* 3 is due to secretion of antibiotics. *Antimicrobial Agents Chemotherapy,* 45: 3156–3161.

Rolfe R.D. 2000. The role of probiotic cultures in the control of gastrointestinal health. *Journal of Nutrition,* 130: 396-402.

Parvez S., Malik K.A., Ah Kang S. and. Kim H. -Y 2006. Probiotics and their fermented food products are beneficial for health. *Journal of Applied Microbiology,* 100: 1171–1185.

Salminen S., Isolauri E., Salminen E. 1996. Clinical uses of probiotics for stabilising gut mucosal barrier: successful strains and future challenges. *Antonie Van Leeuwenhoek,* 70: 347-58.

Saggioro A. 2004. Probiotics in the treatment of irritable bowel syndrome. *Journal of Clinical Gastroenterology,* 38: S104–S106.

Schillinger U. and Lucke F.K. 1987. Identification of lactobacilli from meat and meat products. *Food Microbiology,* 4: 199-208.

Schultz M., Timmer A., Herfarth H.H., Sartor R.B., Vanderhoof J.A., Rath H.C. 2004. *Lactobacillus* GG in inducing and maintaining remission of Crohn's disease. *BMC Gastroenterology,* 4: 5.

Schultz M. and Sartor R.B. 2000 Probiotics and inflammatory bowel diseases. *American Journal of Gastroenterology,* 95: 19S–21S.

Shahani K.M. and Ayebo A.D. 1980. Role of dietary lactobacilli in gastrointestinal microecology. Proceedings of the VI International Symposium on Intestinal Microecology. *American Journal of Clinical Nutrition,* 33: 2448–2457.

Sneath P.H.A., Mair N.S., Sharpe M.E. and Holt J.G. 1986. *Bergey's Manual of Systematic Bacteriology.* Vol 2. Williams and Wilkins, Baltimore, 1599p.

Steer T. 2000. Perspectives on the role of the human gut microbiota and its modulation by pro- and prebiotics. *Nutrition Research Reviews,* 13: 229-54.

Benchimol E.I. and Mack D.R. 2004. Probiotics in the treatment and prevention of acute infectious diarrhea in infants and children: a systematic review of published randomized, double-blind, placebo-controlled trials. *Journal Pediatrics Gastroenterology and Nutrition,* 33: S17–S25.

Van Niel C.W., Feudtner C., Garrison M.M., Christakis D.A. 2002. Lactobacillus therapy for acute infectious diarrhea in children: a meta-analysis. *Pediatrics,* 109: 678–684.

Vanderhoof J.A. 2000. Probiotics and intestinal inflammatory disorders in infants and children. *Journal Pediatrics Gastroenterology and Nutrition,* 30: S34–S38.

Vanderhoof J.A. 2001. Probiotics: future directions. *American Journal of Clinical Nutrition,* 73: S 1152– S 1155.

Walker W.A. 2000. Role of nutrients and bacterial colonization in the development of intestinal host defense. *Journal Pediatrics Gastroenterology and Nutrition,* 30: S2–S7.

Wan-Kyu Lee, Hyeong-Jun Lim, So-Young Kim, Hiromi Kimoto, Sadahiro Ohmomo, Yashuhito Tashiro and Hideki Takebe. 2004 Hypocholesterolemic Effect of *Lactococcus lactis* subsp. *lactis* biovar *diacetylactis* N7 and *Lactococcus lactis* subsp. *lactis* 527 Strians in SD Rats. *Bioscience Microflora,* 24 (1): 11-16, 2004.

Wells J.M., Robinson K., Chamberlain L.M., Schofield K.M. and Le Page R.W. 1996. Lactic acid bacteria as vaccine delivery vehicles. *Antonie van Leeuwenhoek,* 70: 317-30.

Wells J.M., Wilson P.W., Norton P.M., Gasson M.J. and Le Page R.W. 1993. Lactococcus lactis: high level expression of tetanus toxin in fragment C and protection against lethal challenge. *Molecular Microbiology,* 8: 1155-1162.

Xiao S.D., Zhang de Z., Lu H. 2003. Multicenter, randomized, controlled trial of heat-killed *Lactobacillus acidophilus* LB in patients with chronic diarrhea. *Advqnce Therapeutics,* 20: 253–260.

20

Medicinal Plants: Cultivation and Conservation in Northeast India

Nilakshee Devi

1. Introduction

India is one of the centres of Mega biodiversity countries of the world and is bestowed with wild natural treasure. Indian flora comprises of invaluable sources of medicinal and aromatic plants represented by 8,000 species, out of which 386 families and 2,200 genera include families like Asteraceae, Euphorbiacae, Lamiaceae, Fabaceae, Rubiaceae, Poaceae, Acanthaceae and Rosaceae. Out of 17,000 plants, the classic systems of medicines like Ayurveda, Siddha and Unani make use of about 2,000 plants in various formulations.

Northeast India, being a part of Eastern Himalayas, represents one of the "Hot spot" areas of biodiversity. The entire Northeastern Region has rich diversity of medicinal plants (about 400 species). These plants are getting commercial significance during the last few decades because of the popularity of herbal drugs, herbal cosmetics and nutraceuticals. The recent trend of increased unsustainable exploitation from the wild has simultaneously led to the shrinking of medicinal plants in Northeastern Region. Overexploitation has been one of the impacts of market forces, which in turn has made many plant species endangered in this region.

Therefore, conservation and sustainable use of medicinal plants are prime issues in the present context of conserving biodiversity. Cultivating and maintaining plants of medicinal value in herbal gardens, parks and other suitable sites can accomplish ex-situ conservation of medicinal plants. The suitable agro-climatic conditions of the area ensures immense possibilities for establishing medicinal plant gardens in Northeast India with little efforts and this in turn will provide potential return to the farmers.

2. Medicinal Plant Conservation—Indian Scenario

The National Medicinal Plant Board (NMPB) has identified and prioritized 32 medicinal plants for cultivation at the national level, which includes *Emblica officinalis, Saraka asoka, Eithenia somnifera, Aconitum heterophyllum, Aegle marmelos, Phyllanthus amarus, Bacopa monneiri, Santalum album, Swertia chirata, Berberis aristata, Tinospora cordifolia, Gymnema sylvestre, Commiphora wightii, Plantago ovata, Nardostachys jatamanshi, Gloriosa superba, Andrographis paniculata, Crocus sativus, Garcinia indica, Saussaria costus, Picrorhiza kurroa, Solanum nigrum, Glycorrhiza glabra, Chlorophytum borivillianum, Coleus barbatus, Piper longum, Rauvolfia, serpentina, Cassia anguistifolia, Asparagus recemosus, Ocimum sanctum, Embelia ribes* and *Aconitum ferox.*

The Government of India through the Ministry of Commerce has prohibited export of the following medicinal plants under Notification No. 24 (RE-98)/ 1997-2002, New Delhi, dated the 14.10.98.

1. Beddomes cycad *(Cycas beddomei)*
2. Blue Vanda *(Vanda coerulea)*
3. *Saussurea costus*
4. Ladies slipper orchids *(Paphiopedilium sp)*
5. Pitcher plant *(Nepenthes khasiana)*
6. Red Vanda *(Renathera imschootiana)*
7. *Rauvolfia serpentina* (Sarpagandha)
8. *Ceropegia* species
9. *Frerea indica* (Shindal Mankundi)
10. *Podophyllum hexandrum* (emodi) (Indian Podophyllum)
11. *Cyatheaceae* species (Tree Ferns)
12. *Cycadaceae* species
13. *Dioscorea deltoidea* (Elephant's foot)
14. *Euphorbia* species (Euphorbias)
15. *Orchidaceae* species (Orchids)
16. *Pterocarpus santalinus* (Red Sanders)
17. *Taxus willichiana* (Common yew or Birmi leaves
18. *Aquilaria malaccensis* (Agar wood)
19. *Aconitum* species

20. *Coptis teeta*
21. *Coscinium fenestratum* (Calumba wood)
22. *Dactylorhiza hatagirea*
23. *Gentiana*
24. *Gnetum* species
25. *Kampheria galenga*
26. *Panax pseudoginseng*
27. *Picrorhiza kurrooa* Kurroo (Kuru, Kutki)
28. *Swertia chirata* (Charayatah)

India is one of the major exporters of crude drugs mainly to the six developed countries, viz. USA, Germany, France, Switzerland, UK and Japan who share among them 75 to 80% of the total export of crude drugs from India.

Item	**1996-97**	**1997-98**	**1998-99**
Crude drugs	2559501617	2831380625	2824862157
Ayurvedic not put for retail	395000210	422417579	406265096
Ayurvedic put-up for retail	985250139	1187745520	1232053813
Total	**3939751966**	**4441543724**	**4463181066**

Source: 'CHEMEXCIL. *Mumbai.*

Table 1
India's exports of plants or plant parts of medical use

Sl. No	**Plant parts**	**Items**	**Rs. in Lakhs**
01	Ginseng Root		802.45
02	Rhizomes	Netherlands	10.17
		Saudi Arabia	5.72
03	Other Ginseng Root	Australia	11.33
		Canada	104
		Denmark	18.04
		France	21.01
		German F. Rep	113.08

(Contd...)

3. Cultivation of Medicinal and Aromatic Plants (Map) in Northeast India

It is clear that in Northeastern Region there is a Potential Market of Medicinal and Aromatic Plants. Out of about 400 species found in the Northeastern Region, 30 plant species have good commercial demand. The present paper highlights some important aspects of sustainable cultivation of certain common medicinal and aromatic plants available in Northeast India. Some of the potential plants in this category are as under:

A. Cultivation of Medicinal Plants growing at High Altitude

1. *Aconitum species* (Ranunculaceae)

The species of *Aconitum* are commonly known as "Vish" which has been recognized as important medicinal plants both in Ayurveda and Unani Systems of medicine. They are commercially called as Aconites. They are perennial herbs growing wild in the Alpine Himalayas at an altitude of about 3000 - 4000 m.

Tuberous roots of *A. heterophyllum* Wall are known as Atis or Patis and *A. chasmanthum* as Vish or Vatsanabha. *A. ferox* Wall ex Ser is known as 'Indian Aconite'. The Red Data Book of India included *A. ferox* and *A. deinorrhizum* as endangered plants.

The tincture of aconites has been used as heart and nervine sedative. The roots contain alkaloids viz. hetrophylsive, hetrophylline, hetrophyllidine, heteratisine, artisine, altidine, F-dihydroatisine, Isatisine, Hetisine, Hetidines and Hetisinone. The alkaloid content varies from 3 to 4.9%.

Aconites can be cultivated at high altitude. They require cool climate and well-drained loamy soil. Propagation can be done from seeds and tuberous roots. The roots are dug up after the stems have died down. The smaller roots are used for propagation where as the larger ones are used for drugs. If seeds are used, they are sown soon after harvesting.

Economic values of Aconites

It is estimated that the present demand of Aconites in India is about 30 tons per year. The market rate is about Rs.600/- per kg.

2. *Chlorophytum borivillianum* Sant (Liliaceae)

Common name: Safed Mushli

It is a high altitudinal plant. It is an important medicinal plant of Ayurvedic, Allophathic, Homoepathic and Unani system of medicines. The plant is highly demanded for its rich source of protein. It is a well-known tonic and aphrodiasic drug given to cure general debility. Active ingredients present in Safed Mushli are Carbohydrates 10 - 42%, Protein 0.5%, Saponin 2-4% and root fiber 3-5%.

The plant is propagated by seeds and tuberous roots and planted in the second week of June. Plantation is immediately followed by first irrigation. If there is no rainfall, then irrigation may be done after 10-15 days interval.

Harvested roots are spread in the shade for about 4-7 days. Later, fleshy roots are separated from bunches. It has large and consistent market demand in the country and current projection of the annual demand is estimated between 300 - 500 tons.

Economic values

Cost of cultivation = Rs. 22, 000/- ; Gross return = Rs. 65,000/-

3. ***Picrorrhoza kurroa*** Royle ex Benth (Scrophulariaceae)

The plant is found in the Himalayas, from Kashmir to Sikkim at an elevation of 2,700-4,500 m. It is a perennial creeping herb, which spreads by stolons. However, from the wild sources the total annual production is about 2,500 metric tons mainly from Nepal and the states of Himachal Pradesh and Assam.

Other names for *Picrorhiza kurroa* include: Hellbore, Katuka, Katuki, Kurri, Kutki, Katuko, Kuru and Katukarogani. *Picrorhiza kurroa* is a herbal medicine used to treat liver problems, asthma and protection of the liver. The roots and rhizomes are used as a hepatoprotective, antiasthmatic, immunomodulatory and anti-inflammatory.

Cultivation is done in two different seasons using fresh rhizomes as propagule. Crop performance is better in polyhouse than in open field in terms of height, leaf-size and growth of stolons. Autumn season (mid September to mid October) is favourable for establishment of the transplants and production of viable seeds over summer season (April-May). The crop remains dormant in field during January-February but it grows profusely in polyhouse. The crop in open field is also adversely affected due to heavy rains during July-August.

Seeds sown in upper soil surface in Styrofoam seedling trays and covered with thin layer of moss increase the germination capacity. Seeds are sown during November-December in polyhouse at lower altitude, during March-April in nursery beds at middle altitude (2,200m) and during May in alpine area. Seedlings raised at lower altitude are grown in nursery beds for at least 6 months during winter and are transplanted to higher altitude during spring. Vegetative propagation is done successfully through stolons.

The roots and rhizomes are harvested during the senescence of aerial parts in the month of September at the lower altitudes and in October at higher altitudes so as to get the maximum bioactive compounds.

Economic values

The average yield is 450 kg/ha. However, from high dose of forest litter treated field, maximum yield of 612 kg/ha can be obtained. The rate of rhizome/roots ranges from Rs.100 to Rs150 per kg. The leading ayurvedic firms in India are purchasing the crude drug at the rate of Rs.150 per kg. of dried herb.

B. Cultivation of Medicinal and Aromatic Plants in Plain Area

1. ***Aleo vera*** Tourn ex Linn (Liliaceae)

Ass: Ghrito kumari, Salkumari; Eng: Lily of the desert

The plant is a dwarf, aromatic and succulent herb growing in semi-wild state in most parts of the country including Himalayan region. In Indian system of medicines, *Aloe* is popularly known as "Mosabbar". The whole plant including dried juice pulp is medicinal. It regulates blood pressure and acts in rheumatism, arthritis, kidney infections and the urinary tract infection. It acts as a source of nourishment for HIV patients and stimulates immune response against cancer. It is commercially used in cosmetic, shampoo and antidandruff cream. Its bitter property is used in beverages. Aloe based soft drink is popular in Europe. Active constituent of aloe leaves is a mixture of glycosides aloin and barbalion.

The plant can be easily cultivated in almost all parts of India. Well-drained loamy soil is preferable for growth. Propagation can be done by suckers or by cuttings of new shoot or by seeds. The plants need repotting every 2-3 years. Total production of leaf juice of *Aloe* in India is estimated to about 1,00,000 tons.

Economic values

The price of dried Aloes in India ranges from Rs.600 to Rs.1,000 per kg. depending upon the aloin content and colour of the dried leaves. However, the rate of fresh *Aloe* leaves varies from Rs.40 to Rs.55 per kg.

2. ***Andrographis paniculata*** **(Burm.f.) Wall. *ex* Nees (Acanthaceae)**

Mahatita, Kalpatita, Kalmegh (Ass); Dawai wieh (Khasi); The king of bitter (Eng).

A common herb, found in tropical moist deciduous forests, is mostly cultivated in homestead almost throughout Northeast. The whole plant is bitter and medicinal. It is a common indigenous medicinal plant. Stem, leaf, flower, seed and root are all medicinal. It is a good tonic for irregular bowel, loss of appetite, chronic malaria, jaundice and liver complaint. The plant is febrifuge, stomachic, alterative and anthelmintic. Kalmeghin, Andrograpolide (0.8-2%) are two main active compounds. Andrographolides were found to have liver protective properties.

In cultivation, it is raised from seeds. Seeds are sown during May-June. The seedlings are then transplanted in a bed of 60 x 30 cm. 2-3 irrigations may be required during the dry season. It flowers during August to November. The plant is harvested during February to March when the whole plants starts maturing. The whole plant is dried in shades before selling.

Economic values

The average rate in Indian market in Rs.1,800 per quintal. However, the local collectors paid only Rs.6 per kg of the dry herb.

Cost of cultivation = Rs.10,000/- per ha; Gross return = Rs.43,000/- per ha and Net return = Rs.33,000/- per ha.

3. ***Asparagus racemosus*** Wild (Liliaceaea)

Ass: Satmul

An extensive scandent, much branched, spinous twining plant, it is distributed in tropical and sub-tropical parts of India up to an altitude of 1,500m. The plant is diuretic, cardiac and nervine tonic. It is reported to improve intellect, used in treatment of asthma, insanity and epilepsy. It is a potent nervous tonic and is anti-anxiety agent. It is considered to be good for the heart. The tuberous roots of *A. racemosus* constitute the "Shatavar" of Ayurvedic system of medicine. The roots are extensively utilized as nutritive tonic and given for sexual debility. Roots contain steroidal glycosides known as Satavarins.

The plant can be successfully propagated from adventitious roots, which are dipped in liquid of cow dung for 24 hours before planting in the bed during March to April. The sprouted saplings are later transplanted during July to August. The plants require hoeing and weeding during rainy season. The field should be free from water logging. Application of nitrogenous manure is beneficial. The inflorescence appear during August to November. The *Asparagus* plant fruits during December-January. The roots are borne in compact and are fleshy and spindle shaped. They are marketed in pieces of 5-15 cm long and 2 cm thick.

Crude drug: Peeled, dried un-branched tubers, 5 to 15 cm long and up to 1 cm broad at the thickest part, almost straight, fusiform and somewhat translucent.

Fresh drug: The tubers are generally deeply wrinkled longitudinally. The wrinkles appear as flaps extending to the whole length of the root. The surface is smooth, waxy, having fawn or flesh colour.

4. ***Cymbopogon flaxuosus*** (Poaceae)

Common name: Lemon grass

It is a perennial grass and naturally occurring plant. It is cultivated in northern plains, southern plains and northeastern states of India at commercial scale. Plants are cultivated for production of essential oil.

Slips obtained from 4-5 year old plantations are used as planting material. In southern states seeds are used as planting material. The crop flowers in November to December and produces viable seeds in January to February. Seeds from one year old plants are preferred for raising plants in nursery. The optimum time of planting the slips is the onset of monsoon rains. Water logged areas are not suitable for this type of grass. The first harvest is ready after 90 days. 3 harvests can be obtained in the first year and 4 to 5 harvests during each of the 3 to 5 succeeding years. The crop is cut after every 60 to 65 days, except during the dry season.

Economic values

The herbage contains 0.35% oil on fresh weight basis. The grass yield ranges from 18 to 25 tons per hectare, giving 60 - 72 kg of oil.

5. ***Rauvolfia serpentina* Benth ex Kurz.** (Apocynaceae)

Devil's pepper (Eng) Sarpagandha (Ass) Tdong-pait-puraw (Khasi)

It is an evergreen, perennial indigenous herb found widely distributed in the Himalayan foothills and Peninsular India. It is popularly known as "Pagal ki Dova" in Indigenous System of Medicine. Roots are traditionally used in insomnia, insanity, asthma and high blood pressure. Roots are officinal in Indian pharmacopoeia as anthelminitic, appetizer and sedative. Roots are the source of "Reserpine". Roots of *Rauvolfia* and its products command a sizeable market both in India and abroad.

Rauvolfia grows in sandy-loam to clayey-loam acidic soils, rich in humus. It prefers tropical to subtropical climate, receiving heavy monsoon rains between June and August and the temperature between 10° and 38°C. It is a 1 $^{1}/_{2}$ to 2 years crop and can be grown from seeds, root, root-shoot and shoot-cuttings. Seeds are sown in well prepared seed beds in lines of 20cm apart, during April and May. When the seedlings grow 5 to 8 cm tall in the next 2 months, they are transplanted in well prepared field. The crop requires 5 weeding and hoeing. The roots are dug out in November to January in northern India and February in southern India. The root bark of the thin fibrous roots contain a larger percentage of alkaloids and should not be damaged a loss during digging. The roots are dried in the sun and in the process 55-60 % of their weight is lost. The average yield is 950 kg. per hectare.

Economic values

Cost of cultivation per hectare = Rs.19,000/-, Return = Rs.60,000/- and Net income Rs.41,000/-

6. ***Tinospora cordifolia* Miers** (Menispermaceae)

Ass: Hagunilota/Sidhi lota ; Eng: Giloy, Guruch, Indian Quinine

The climber is found throughout the tropical regions of North and South India extending from Kumaon to Assam, up to 1200 m elevation above mean sea level. The whole plant i.e. stem, root, leaves and fruits have medicinal properties. The starch obtained from dry stem is used for high fever, urinary trouble, abdominal pain, gout, gonorrhoea and tuberculosis. Fruits are used in jaundice and rheumatic pain. Roots are used in leprosy. The drug "Guduchi" or "Amrita" derived from the stem of *T. cordifolia* is a well-known Indian bitter and "Rasayan-drug of the Ayurveda". The plant is prescribed for the treatment of fever, diabetes, jaundice etc.

Active Constituents: Bitter Principle Columbin, Palmarin, Tinosporon, Tinosporic acid and Tonosporol.

It grows well in almost all types of soil under varying climatic conditions. The plant is cultivated by stem cutting in the month of May-June. It requires some support preferably Neem and Mango trees. Nursery of supporting trees should be prepared before planting of Giloy. After one year of planting of supporting plants, cuttings of Giloy are planted near these trees. Periodical hoeing is done, both in the nursery and field as per requirement. The plant matures in 8-10 months. Mature plants are collected when the stem becomes 8-10 cm thick and the leaves start to fall. They are cut into small pieces and dried in shade.

Economic values

Approximately 10-12 quintal of dried stem can be obtained from 100 kg of dried stem. Farmers may earn extra profit from supporting trees.

7. ***Withania somnifera* (L.) Dunal** (Solanaceae)

Common name: Ashwagandha

The plant is also known as Indian ginseng, the roots of which have been employed in Indian traditional systems of medicine. It grows in dry and sub-tropical regions up to an altitude of 1,500m. The estimated production of Ashwagandha roots in India is more than 1,500 tons and the annual requirement about 7,000 tons. Ashwagandha is considered to be one of the best rejuvenating agents in Ayurveda. Root drug is used in treatment of rheumatic pain, inflammation of joints, nervous disorders and epilepsy, as tonic for cold, cough, female disorders, as a sedative, in care of senile debility, ulcers etc. Leaf juice is useful in conjunctivitis. Bark decoction is taken for asthma. The plant and its extracts are used in preparation of herbal tea, powders, tablets and syrups.

Active Constituents: Somniferine, somnine, somniferinine, withananine etc. Alkaloid percentage in roots ranges from 0.13 to 0.31%.

Ashwagandha grows well in sandy loam or light red soil having pH 7.5 to 8.0 with good drainage system. It is grown as late rainy season (kharif) crop. The crop requires relatively dry season during its growing period. It can tolerate a wide range of temperature. It is propagated by seeds. Fresh seeds are sown in well prepared nursery beds. Transplantation is preferred for better quality plants needed for export purpose. For this, a well maintained nursery is prerequisite. Nursery is raised in the month of June-July. Seeds are sown just before the onset of monsoon. Seedlings about 35 days old are transplanted in the main field. The crop is harvested for roots in January to March i.e. 150 to 180 days after sowing. Roots are dug out or ploughed using power tiller or a country plough. The tap root should be carefully taken out not to damage even the small lateral roots.

The roots are separated from the aerial portion by cutting the stem 1 to 2 cm above the ground. Better root pieces can be graded in the following grades as per its length and thickness.

i. **A grade root**: Root pieces up to 7cm and diameter 1.0 – 1.5 cm, solid, bright and pure white.

ii. **B grade root**: Root pieces up to 5 cm and diameter 1cm, bright and white.

iii. **C grade root**: Root pieces up to 3-4 cm and diameter less than 1 cm, solid with side branches.

iv. **Lower Grade**: Small root pieces, semi-solid, very thick, yellowish, and chopped. Berries are hand plucked separately. They are dried and crushed to take out the seeds.

Economic values

Yield: 3 to 5 quintal of dry roots per hectare; 50 to 75 kg of seeds per hectare

Income: From roots- Rs. 35,000/- per ha; from seeds Rs 3,600/- per ha;

Gross income = Rs 38,600/ha; Cost = Rs 22,400/ha; Net income = Rs 16,200/ha

C. Cultivation of Medicinal and Aromatic Plants in Waterlogged Conditions

About 11 lakh hectares of land in our country is affected by the problems of water logging. Northeast India, especially Assam has to face this situation due to flood. Most of the land remains unutilized because the productions of traditional crops in these areas are risky. Some medicinal and aromatic plants can be grown in these areas more profitably with lesser risk.

1. ***Acorus calamus*** **Linn** (Araceae)

Bosh (Ass); Sweet Flag (Eng); Khasi: Bat–Bhut, Krah phang Phiang (Dkhian), Simprabet (Nongstoin), Bet (Mawsynram)

It is a semi-aquatic aromatic perennial herb; this plant is found in marshy area and rice field. The plant is distributed in the Himalayan region up to a height of about 6,000 ft and is common in entire Northeastern Region. Dry rhizome is aromatic and medicinal, being used for the treatment of gastric, respiratory, cough and cold related problems. The root is a very good tonic for asthma, remittent fever. Leaf juice is used in beer, gin, vinegar etc. The plant is a good insecticide and snake repellent. Rhizome contains 1.5 to 3.5 % yellow aromatic oil.

This plant is grown as perennial crop. The crop is grown in clayey loams of riverside. Rhizome along with 2-3 leaves is used as planting material. It is planted during the month of February to March in waterlogged regions, so that a good mass can be generated before water logging condition starts. 3-4 weeding may be required after recession of the flooded water. Hoeing should be avoided. After two years, digging of suckers is done in the month of January – February. Suckers are dried in shade and kept in gunny bags. About 25 quintals of dried rhizome can be obtained from two years old crop. The rhizome can be stored for a long period.

Economic value

If dried rhizomes are sold @ Rs. 20/- per kg, then net profit will be Rs. 25,000/- per ha.

2. ***Bacopa monneieri*** (Scrophulariaceae)

Ass: Brahmi

A small herb found throughout India up to 4,000 feet. The plant is a creeper, mostly found near waterlogged place. The dried herb is valued for its use as tonic and memory enhancer in the traditional system of medicine. Major chemical compounds present are- Bacoside A and Bacoside B.

Cultivation: 7-10 cm long herbs are selected as planting materials. It is also propagated by runners and by seeds in rainy season. Planting should be done during the month of February to March in water logged regions. 2-3 weeding may be required after recession of the flooded water.

Economic values

About 20 quintal of dry herb can be obtained from 2-3 cuttings in one year.

Cost of cultivation = around Rs. 20,000/- per ha. If dry herbs are sold @ Rs. 20/Kg, then net profit will be Rs. 20,000/- per ha. Brahmi has a good market especially for its "Brahmi oil" due to its high medicinal value. Rate of dried drug in retail market varies from Rs.20-25/- per kg.

3. *Vitiveria zizanioides* (Poaceae)

It is an aromatic herb. Essential oil obtained from the roots of this plant is used in valued perfumes and cosmetic materials. It is also used in different drinks.

Propagation of *Vitiveria* is done by the slips obtained from old plantation. Transplantation should be done in February – March in the areas affected by water logging. The crop can be dug after one or two years for the production of roots and essential oil. About 15 quintal of semi-dried roots and 10-12 kg oil can be obtained from two years old crop whereas 6-7 quintal of semi-dried roots and 4-5 kg. oil can be obtained from one year old crop.

Table 3

Common/ trade name	Scientific name	Duration (Year)	Input Cost (Rs)	Output Cost (Rs)
Amla	*Emblica officinalis*	After 6	15,000	50,000
Atis	*Aconitum heterophyllum*	2	60,500	1,25,000
Ashok	*Saraca asoka*	10	5,500	20,000
Ashwagandha	*Withania somnifera*	1	6,000	19,000
Bach	*Acorus calamus*	1	25,000	60,000
Brahmi	*Bacopa monnieri*	1	15,000	50,000
Tulsi	*Ocimum sanctum*	1	1,500	5,000
Bhui amlaki	*Phyllanthus amarus*	1	2,000	4,000
Guduchi	*Tinospora cordifolia*	1	2,500	6,000
Kalmegh	*Andrographis paniculata*	1	8,000	12,000
Makoy (Lunilunika)	*Solanum nigrum*	1	5,500	18,000
Chirayita (Himalayan)	*Swertia chirata*	1	76,000	1,25,000
Chandan	*Santalum album*	After 30	75,000	4,50,000

Cost of cultivation = around Rs. 60,000/- per ha for two years crop; around Rs. 40,000/- per ha for one year crop. Net profit = Rs 60,000/- per ha from two years crop; Rs. 20,000 – 25,000/- per ha from one year crop, if oil price is around Rs. 10,000/- per kg.

Economic Importance of some Medicinal Plants (Per Acre) 2004

National Medicinal Plants Board has emphasized on cultivation economics of certain medicinal plants. Table 3 includes names of some medicinal and aromatic plants common to North Eastern region.

4. Conclusion

"Medicinal plants are exhaustible if over exploited and sustainable if used with care and wisdom". Cultivation of medicinal and aromatic plants can promote and maintain the health of local communities, besides generating productive employment. Systematic cultivation of these plants and their drying/ preservation at farm level constitutes a promising rural industry with attractive remunerations. These semi-processed plants find a ready market with herb dealers/large pharmaceuticals industries. To save these medicinal and aromatic plants, they should be cultivated commercially and collection from wild sources should be completely banned, so that in future these plants will be available in the market for preparation of drugs.

REFERENCES

Report of the Task Force on Conservation and Sustainable Use of Medicinal Plants (March-2000), Government of India, Planning Commission.

Wealth of India: Raw Materials (1985), Vol 1:A. Publications and Information Directorate, CSIR, Hill side Road, New Delhi – 110012.

MAPs Dew (2007), Vol 3 (1-2). Director, Central Institute of Medicinal and Aromatic Plants, Lucknow – 226015.

Index

■■■

Gaultheria fragrantissima

Solanum khasianum

Litsea citrata

Swertia chirata

Potentilla fulgens

Piper longum

Pogostamon cablin

Taxus baccata

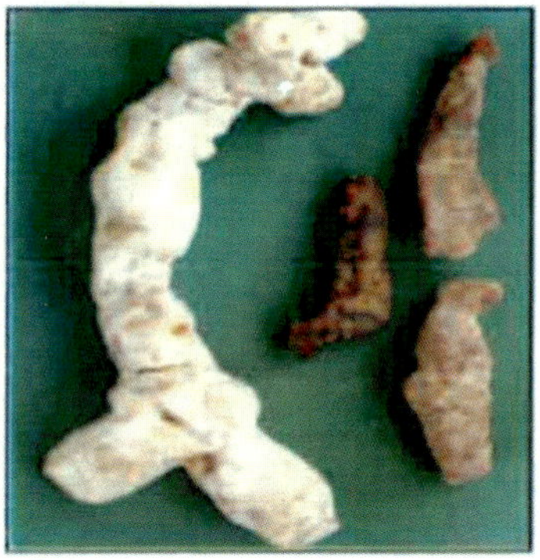

Smilax glabra rhizomes

Flemingia vestita

Curcuma longa

Acorus calamus

Adiantum phillipense

Averrhoa carambola

Begonia sp

Bauhinia sp

Cinamomum tamala

Costus speciosus

Garcinia sp

Fruits Tejmui

Citrus latipes

Curcuma sp

Z. hamiltonianum (Tejmui)

Closed pitcher of N. khasiana

Opened pitcher

N. khasiona growing in Tura (Garo Hills)

(a) Osbeckia stellata

(b) Flowers of O.stellata

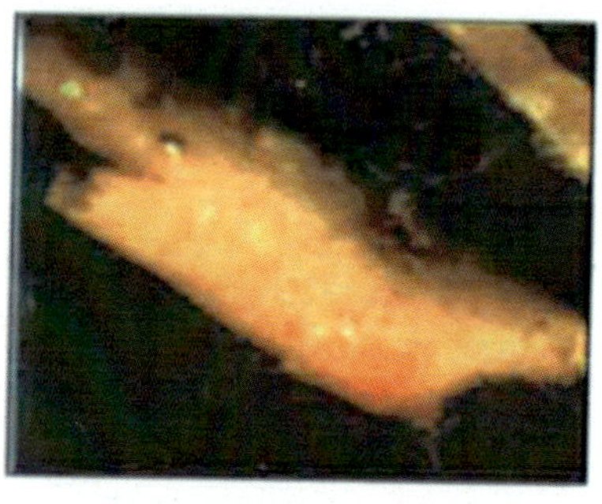

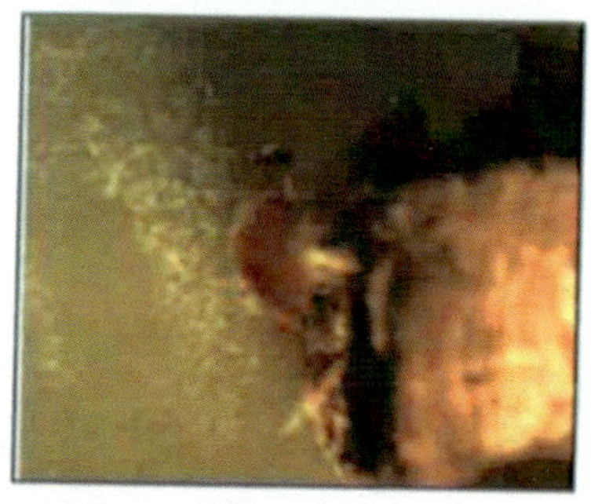

(a) & (b) Fungal hypahe seen emereging out of the plant materials (4 mX magn.)

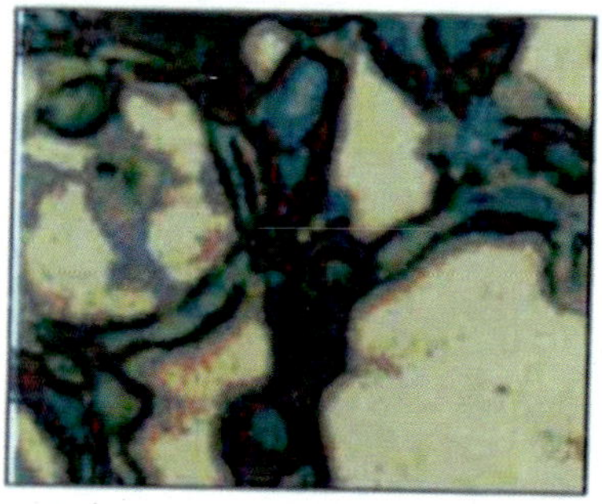

Fungal hyphal structure (40 X magn.)

Bistorta major

Fagopyrum esculentum

Fagopyrum dibotrys

Persicaria barbata

Persicaria hydropiper

Persicaria chinensis

Persicaria lapathifolia

Persicaria orientalis

Persicaria perfoliata

Persicaria micrecephala

Poygonum aviculare

Polygonum plebeium

Polygonum molle

Rumex maritimus L.

Rumex nepalensis

Rumer vesicarius

Abelmoschus esculentus

Allium cepa

Allium sativum

Curcuma domestica

Basella alba

Mentha spicata

Centella asiatica

Piper nigrum

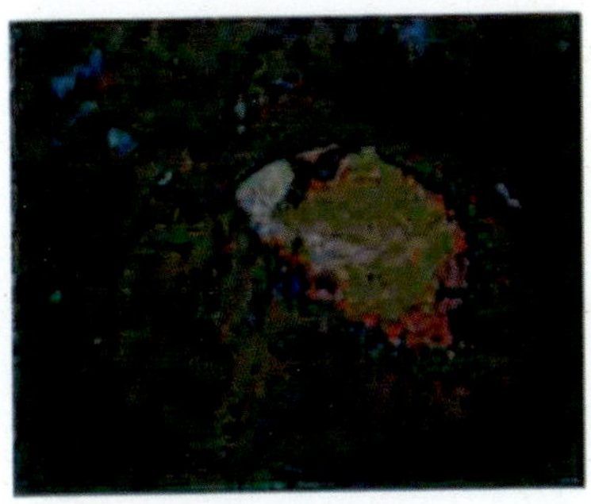

Feronia limonia

Lantana camara

Aloe vera

Artocarpus heterophyllus

Coleus aromaticus

Mimosa pudica

Musa paradisiaca

Saccharum officinarum

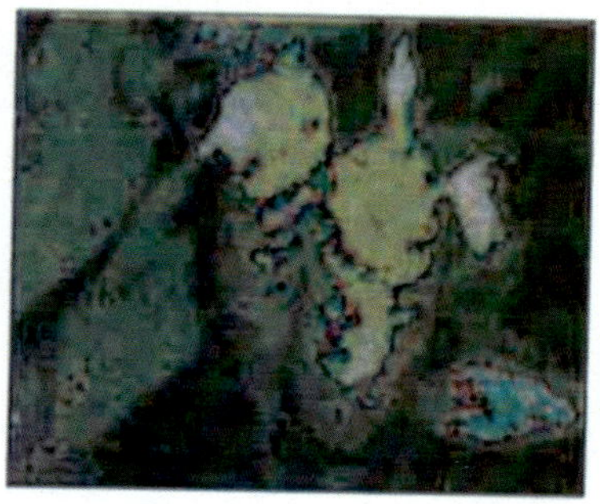

Zingiber zerumbet

Punica granatum

Phyllanthus emblica

Eupatorium adenophorum